Viruses
Plant Disease

Viroids and
Viroid Diseases

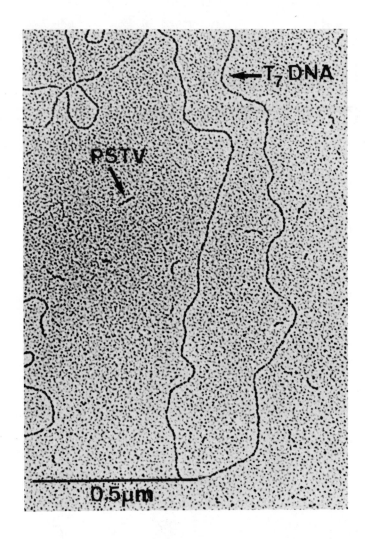

Electron micrograph of a viroid (PSTV) and a viral DNA (coliphage T_7 DNA), illustrating the smallness of the viroid as compared with the viral nucleic acid. Courtesy: Th. Koller and J. M. Sogo, Swiss Federal Institute of Technology, Zurich, Switzerland.

VIROIDS AND VIROID DISEASES

T. O. DIENER

4530 Powder Mill Road
Beltsville, Maryland

A Wiley-Interscience Publication

JOHN WILEY & SONS
New York • Chichester • Brisbane • Toronto

Library of Congress Cataloging in Publication Data

Diener, Theodor Otto, 1921–
  Viroids and viroid diseases.

  "A Wiley-Interscience publication."
  Bibliography: p.
  Includes index.
  1. Viroids. 2. Viroid diseases of plants. I. Title.

QR500.D53   576'.64   78-21681
ISBN 0-471-03504-1

Printed in the United States of America

10 9 8 7 6 5 4 3 2 1

To Dr. Samuel Blumer, whose example, friendship, and encouragement inspired me to embark on a research career, and to Sybil, my wife, who had the fortitude to live with me during conception and execution of this book.

PREFACE

Viroids are the smallest known agents of infectious disease. They are responsible for a number of destructive diseases of cultivated plants, but may also occur and cause disease in animals. Although some of the diseases that viroids cause had been known for decades, these diseases were generally believed to be due to infection by conventional viruses. The unique nature of their causative agents came to light only in 1971, when the agent of the potato spindle tuber disease was shown to be a low molecular weight ribonucleic acid (RNA) with unusual properties that is able, despite its small size, to replicate autonomously in susceptible cells. Recognition of the profound disparity between this pathogen and all other pathogens, including all known viruses, led to the proposal to call such agents *viroids*.

Since then, several other plant diseases have been shown to be caused by similar low molecular weight RNAs, and due to the efforts of a number of investigators much has been learned of the unique physical-chemical and biological properties of viroids. This book endeavors to present a comprehensive and up-to-date account of our knowledge of viroids and of the diseases they cause. Also, and possibly more importantly, the book attempts to point out the many areas of ignorance in our understanding of viroids as biological entities. We do not know, for example, how widespread viroids are in nature, how serious a potential threat they are to agriculture, whether they occur in forms of life other than higher plants, how they replicate, and how they bring about the metabolic aberrations in some of their hosts that lead to disease and, in some cases, death of the infected plant.

Each of these areas of ignorance affords attractive opportunities to advance our knowledge, not only of viroids and of the diseases they cause, but also of profound and general biological mechanisms. Hopefully, this

book will attract additional investigators into the area of viroid research
and thereby help to accelerate the acquisition of knowledge regarding this
newly recognized group of pathogens.

T. O. DIENER

Beltsville, Maryland
January 1979

ACKNOWLEDGMENTS

I wish to express my indebtedness to all those whose work has contributed so much to our knowledge of viroids and without whose careful recording of observations and experimental results this attempted synthesis could not have been contemplated. I am most grateful to those who permitted me to reproduce their photographs, and I particularly thank Mrs. Margaret "Kitty" Taylor for her tireless help in typing and proofreading the manuscript.

T. O. D.

CONTENTS

Viroids and
Viroid Diseases

1. THE DISCOVERY OF VIROIDS

"S' efforcer de se convaincre soi-même de la vérité qu'on a entrevue est le premier pas vers le progrès; persuader les autres est le second."
Louis Pasteur
C.R. hebd. Séanc. Acad. Sci. Paris **82,** 1079 (1876)

In science, unexpected results of routine experiments are the most frequent source of discovery. Many experiments, however, lead to unexpected results, but almost always these have trivial explanations, such as faulty technique or erroneous suppositions. Significant deviations from theory are rare and, as often as not, are overlooked or blamed on faulty experimentation.

The recognition of viroids as a novel type of pathogen followed this pattern. Efforts to purify the agent of the potato spindle tuber disease, which was generally believed to be a conventional plant virus, led to results that were inconsistent with this belief, as well as with widely held concepts of virology and molecular biology. In this chapter, the prerequisites for the recognition of viroids and the chronology of their discovery are discussed.

1.1 PREREQUISITES FOR DISCOVERY OF VIROIDS

Recognition of the basic disparity between viroids and conventional viruses became possible only after certain basic principles of virology and molecular biology had been established. Foremost among these was the realization that the genetic information of viruses resides in their nucleic acid component, a fact that, in the case of plant viruses, has been most dramatically established with the demonstration that RNA isolated from

1

tobacco mosaic virus is infectious (Gierer and Schramm, 1956; Fraenkel-Conrat, 1956). The isolation of defective tobacco mosaic virus strains from nitrous-acid-treated virus preparations and the demonstration that with these the infectious principle behaves in a manner similar to that of infectious RNA isolated from ordinary virus (Siegel *et al.*, 1962) clearly showed that a virus may be able to persist *in vivo* in the form of free RNA. The discovery of the so-called nonmultiplying or winter forms of tobacco rattle virus and the demonstration that these consist of nonencapsidated viral RNA (Sänger and Brandenburg, 1961; Cadman, 1962) showed that viral pathogens in the form of free RNA occur in nature. Finally, with the demonstration that free, infectious RNA exists also in plants infected with a conventional plant virus (Diener, 1962), the possibility of free RNA viruses occurring naturally became still more plausible. None of these findings, however, suggested the existence of autonomously replicating low molecular weight RNA species or their importance as naturally occurring incitants of damaging diseases of crop plants.

1.2 CHRONOLOGY OF DISCOVERY

In retrospect, it is evident that the recognition of viroids as representatives of a novel class of pathogens that are distinct from viruses required convincing evidence in five crucial areas. It was necessary to demonstrate that:

1. The pathogen exists *in vivo* as a nonencapsidated nucleic acid.
2. Viruslike particles are not detectable in infected tissue.
3. The pathogen is a low molecular weight nucleic acid.
4. The infectious nucleic acid replicates autonomously, that is, without assistance from a helper virus.
5. The infectious nucleic acid consists of one molecular species only.

So far the following viroids have been identified and named after the diseases they cause:

1. Potato spindle tuber viroid (PSTV).
2. Citrus exocortis viroid (CEV).
3. Chrysanthemum stunt viroid (CSV).

4. Chrysanthemum chlorotic mottle viroid (ChCMV).

5. Coconut cadang-cadang viroid (CCCV).

6. Cucumber pale fruit viroid (CPFV).

7. Hop stunt viroid (HSV).

It is interesting to examine the chronology by which evidence in the five crucial areas listed above was reported with each viroid. So far, evidence supporting all five criteria has been reported only for PSTV, whereas evidence with the more recently discovered viroids is still lacking in one or more of the crucial areas.

Here, only the earliest record giving convincing evidence is listed for each viroid; confirming reports and a more detailed discussion of the experimental evidence will be presented in Chapter 4.

1967. On the basis of its low sedimentation rate and sensitivity to ribonuclease, Diener and Raymer (1967) conclude that PSTV is a free RNA. Existence of viruslike particles is considered unlikely because phenol treatment of crude extracts does not change the sedimentation properties of the infectious material, since the latter has the expected buoyant density of RNA, and because the low sedimentation rate cannot be explained on the basis of low density (lipid content) of a putative virion.

1968. The existence of virions in PSTV-infected tissue becomes even less likely when Diener (1968) shows that, *in situ,* PSTV is sensitive to ribonuclease.

With citrus exocortis disease, Semancik and Weathers (1968b) present evidence that the causative agent sediments at a low rate and is sensitive to ribonuclease. Also, no viruslike particles can be identified by electron microscopy in CEV-infected tissue (Semancik and Weathers, 1968a).

1970. Zaitlin and Hariharasubramanian (1970) report that no proteins that can be construed as coat proteins are identifiable in PSTV-infected tissue.

1971. From a combined analysis of PSTV in density gradients and polyacrylamide gels, Diener (1971b) concludes that PSTV is a low molecular weight RNA. In the same report, extensive evidence to indicate that no helper virus is involved in the replication of PSTV is presented. The name *viroid* is proposed as a generic term for PSTV and pathogenic nucleic acids with similar properties (Diener, 1971b).

Diener and Smith (1971) report further evidence for the low molecular weight of PSTV by showing that the RNA is able to penetrate into gels of high polyacrylamide concentration, from which high molecular weight RNAs are excluded.

1972. Because the infectivity dilution curve of PSTV suggests a single-hit event, Diener (1972b) concludes that PSTV consists of a single molecular species.

Semancik and Weathers (1972b), on the basis of its mobility in poly-acrylamide gels, conclude that CEV is a low molecular weight RNA. They show that its buoyant density is that expected of an RNA (Semancik and Weathers, 1972a).

Diener and Lawson (1972) present evidence that the chrysanthemum stunt disease is viroid caused by demonstrating that the infectious agent sediments at a low rate, is sensitive to ribonuclease, and moves with low molecular weight RNAs in polyacrylamide gels.

1973. Semancik *et al.* (1973b) investigate the resistance of CEV to ionizing radiation and report a target volume of about 10^5 daltons, suggesting that CEV consists of a single molecular species.

Hollings and Stone (1973) determine that CSV is highly resistant to ultraviolet light irradiation.

Sogo *et al.* (1973) achieve the first visualization of a viroid (PSTV) by electron microscopy and establish its low molecular weight by direct length measurements.

1974. On the basis of its low rate of sedimentation and sensitivity to ribonuclease, Romaine and Horst (1974) conclude that the agent of the chrysanthemum chlorotic mottle disease is a viroid.

Diener *et al.* (1974) report that PSTV has a very high resistance to inactivation by ultraviolet light irradiation, suggesting that PSTV consists of one molecular species only.

1975. Randles (1975) provides evidence for the association of two RNA species with coconut cadang-cadang disease and suggests that the disease may be viroid incited.

Dickson *et al.* (1975) publish the first RNA fingerprints of viroids (PSTV and CEV) and conclude from their complexity that each viroid is a single, distinct molecular species.

1976. Sänger *et al.* (1976) present evidence that the agent of the cucumber pale fruit disease is a viroid as had been claimed earlier on the basis of unpublished evidence (Van Dorst and Peters, 1974). Evidence presented includes rate of sedimentation, sensitivity to ribonuclease, gel mobility, and length measurements by electron microscopy.

Dickson (1976) presents two-dimensional fingerprint patterns of CSV and the cadang-cadang-associated RNA, and conclude that each is a single distinct, low molecular weight RNA species.

1977. Sasaki and Shikata (1977b) report that the agent of the hop stunt disease sediments at a low rate and is sensitive to ribonuclease. They conclude that the agent is a viroid.

Conejero and Semancik (1977) provide evidence that no protein that has the expected properties of a viral coat protein can be identified in CEV-infected tissue, making it unlikely that viruslike particles exist in such tissue.

Thus, for all viroids so far identified, convincing evidence indicates that they are nonencapsidated RNAs. Some evidence has been presented with all but two viroids (ChCMV and CPFV) that viruslike particles are absent in infected tissue. The low molecular weight of all viroids except ChCMV and HSV has been reasonably well established, but the question of autonomous replication has so far been investigated only with PSTV. Evidence that the viroid is a single molecular species is available for PSTV, CEV, CVS, and CCCV, but not for CPFV, ChCMV, or HSV.

With the elucidation of the unique molecular structure of viroids (see Chapter 7), peculiar structural features, such as circularity and a very high degree of intramolecular complementarity, promise to be useful criteria for the identification of an RNA as a viroid. So far, however, it is not yet certain whether all viroids will prove to possess these structural features.

1.3 DEFINITION OF VIROID

By necessity, the following definition of the term *viroid* is operational, that is, it is based on presently recognized properties of the pathogens:

Viroids are low molecular weight nucleic acids that are present in certain organisms afflicted with specific maladies. Viroids are not detectable in healthy individuals of the same species but, when introduced into such individuals, they replicate autonomously and cause the appearance of the characteristic disease syndrome. Unlike viral nucleic acids, viroids are not encapsidated. Viroids are highly resistant to heat, as well as to ultraviolet and ionizing radiation. They contain extensive regions of intramolecular complementarity and exist as covalently closed circular structures.

Undoubtedly, with increased knowledge of the properties of viroids, this definition will require modification, but in light of present knowledge, it adequately distinguishes viroids from all other pathogens known, including conventional viruses.

1.4 THE TERM VIROID

Although usage of the term *viroid* to denote the pathogens that are the subject of this book is now universal, certain objections to the term were raised at an earlier time. These centered around the following two arguments:

1. Properties of the newly discovered pathogens are not sufficiently different from those of conventional viruses. These pathogens simply constitute very small viruses. No new term, therefore, is needed.

This argument was advanced by McKinney (1973), who regarded viroids as the most primitive or the most degenerate plant viruses known. As has been pointed out (Diener, 1973a), no evidence exists as to whether viroids are or are not related to viruses, and the more recent elucidation of the unique structural properties of viroids (see Chapter 7) makes it unlikely that they evolved from conventional viruses or that they constitute highly degenerate viruses (see Section 11.2). In light of these and other properties of viroids, a new term definitely appears to be needed.

2. A new term is called for, but *Viroid* is not suitable because:
   a. It implies similarity with viruses.
   b. The term has been preempted by earlier usage in different connotations.

Objection (a), which evidently is the opposite of the objection discussed under argument (1), was raised by Semancik *et al.* (1973b), who introduced the term *pathogene* "to provide a position of divergence from viral processes implicit in the terms viroid . . . and metavirus (Hanneman and Singh, 1972)." *Viroid*, however, may be interpreted to indicate similarity of the respective diseases and not necessarily of the respective causative agents. Also, more recently, both Semancik and coworkers and Singh and colleagues have accepted the term viroid (Semancik and Vanderwoude, 1976; Singh et al., 1976).

Objection (b) was raised by McKinney (1973) on the basis that the term still served a useful function in an earlier connotation advanced by Altenburg (1946), in which viroids were considered to be hypothetical, ultramicroscopic organisms that were useful symbionts, occurred universally within cells of larger organisms, and were capable by mutation of giving rise to viruses. As has been pointed out (Diener, 1971b, 1973a), however, Altenburg's "viroid theory" has not been widely accepted, presumably because experimental verification of its principal tenets did not materialize. Also, Altenburg did not coin the term *viroid*, but only redefined it from its earlier meaning, namely, "any prophylactic vaccine" (Stedman, 1961) or "any biological specific used in immunization" (Dorland, 1932). The term, either in Altenburg's or in its earlier connotations, was considered obsolete and redefinition of *viroid* to encompass nucleic acid species with properties similar to those of PSTV appears appropriate and serves a useful function (Diener, 1973a).

2. NATURAL PLANT VIROID DISEASES

Of the naturally occurring plant diseases that are now known to be viroid incited, some have been recognized as distinct maladies by growers and plant pathologists for as long as 50 years or more, whereas others have been known for only a few years. Even though the ultimate cause of these diseases was unknown at the time, systematic observation as well as field and greenhouse experimentation yielded a large body of information on geographical distribution, symptomatology, modes of field transmission and spread, economic losses, occurrence of strains, and other disease aspects. This information is summarized here.

2.1 POTATO SPINDLE TUBER DISEASE

2.1.1 Historical

The spindle tuber disease of potato was first described by Martin (1922), who reported that this "new potato trouble made its appearance in South Jersey in fields of late planted Irish Cobblers grown for seed purposes." Martin noted that the name *spindle tuber* originated with local growers and that in almost every instance affected plants were found in fields planted with potatoes grown in Maine the previous year. Although the author had no positive evidence to the effect that spindle tuber was an infectious disease (such as was known to be the case at that time for potato leaf roll and mosaic disease), he considered this to be likely because by the fall of 1922 some strains of Irish Cobblers that had been grown in New Jersey for a number of years developed the disease (Martin, 1922).

Evidently, the spindle tuber disease was imported into New Jersey from

Maine and in the following year, Schultz and Folsom (1923a–c) reported on their extensive investigations of the disease in Maine. They stated (Schultz and Folsom, 1923a) that the disease had been recognized for many years by growers and others by various names, such as "running out," "running long," "off shape," "poor shape," "reversion," or "senility."[1] The authors described their observations and experimental work from 1917 to 1921 that led to the conclusion that the disease spread in the field, was infectious, and therefore properly belonged to the group of diseases that were at that time known by the term *degeneration disease*. From a historical standpoint, it is interesting to recall that these degeneration diseases had long been blamed on senility, "reversion," or loss of vigor caused by "unnatural" prolonged asexual reproduction of potatoes. At the time of the first report of Schultz and Folsom (1923a), however, the infectious nature of these diseases was recognized and the authors advanced the following definition:

Degeneration diseases of potato are here considered to be those transmissible or infectious diseases which are perpetuated indefinitely by vegetative growth and propagation, and of which no cause, either organic or inorganic, has been identified and demonstrated. . . . For convenience this unknown cause, which seems to be associated with plant juice, will be referred to here, as elsewhere, as a contagium or a virus.

In the same year, potato spindle tuber disease was reported from Vermont (Gilbert, 1923) and further reports on the disease appeared (Fernow, 1923; Folsom, 1923). For a number of years, research on potato spindle tuber disease centered in New England (Schultz and Folsom, 1925; Gilbert, 1925; Bonde, 1927) and Nebraska (Goss and Peltier, 1925; Goss, 1926 a–c; Werner, 1926; Goss, 1928).

2.1.2 Geographic Distribution

The disease is common in the potato-growing regions of northern and northeastern U.S.A. and Canada (Diener and Raymer, 1971). Potato "gothic" virus, described from the U.S.S.R., has been reported to be a combination of "necrotic tuber spot, leaf roll, black peel, and the spindle tuber virus of America" (Leont'eva, 1964). Thus the potato spindle tuber disease appears to be present in the U.S.S.R. It has not been reported

---

[1] Schultz and Folsom (1923a–c) first used the term *spindling tuber disease* but soon adopted Martin's terminology (Folsom, 1923; Schultz and Folsom, 1925).

from Western Europe, but in England the occurrence of potato spindle tuber disease was suspected on the basis of spindle tuberlike symptoms observed in tubers of the cultivar Redskin. Also, original transmission experiments appeared to have been successful (Cammack and Richardson, 1963). Later, however, Cammack (1964) reported that further transmission experiments were unsuccessful and concluded that the abnormality in the cultivar Redskin was not caused by spindle tuber and that there was no longer reason to suspect the presence of the disease in Britain.

Fernandez Valiela and Calderon (1965) reported that the disease occurred in potato growing areas of Argentina.

2.1.3 Symptomatology

Depending on the potato cultivar and on environmental conditions, symptoms of the disease may vary considerably. Foliage symptoms are obscure in many instances and the plants may be severely or not at all stunted. From late spring to midsummer, the foliage often turns slate grey with dull leaf surface. The tubers characteristically are elongated with prominent bud scales ("eyebrows") and sometimes have severe growth cracks (Diener and Raymer, 1971) (Fig. 1).

In their original publications, Schultz and Folsom (1923a, b) noted that in the cultivar Green Mountain the disease was characterized by spindliness and uprightness of potato plants, by more erect and often somewhat darker green leaves than healthy ones, and by slight rugosity of leaves. The tubers were "abnormally spindling, spindle shaped, cylindrical and supplied with conspicuous eyes" (Schultz and Folsom, 1923a).

In Irish Cobbler plants, the disease is manifested by vines that grow upright, branch but little, and are smaller than normal (Martin, 1922). The affected vines die earlier than healthy ones and the leaves are much more narrow and pointed than typical leaves. Tubers from affected plants are long, narrow, smooth skinned, show more eyes than the true type cobblers, and the eyes are sometimes borne on knoblike protuberances (Martin, 1922).

Marginal leafrolling has been reported as a symptom of spindle tuber in New York (Fernow, 1923) and New Jersey (Martin, 1922), but it is not clear whether some of this leafrolling was due to coinfection of plants with leafroll virus.

The effects of environmental factors on potato degeneration diseases, including spindle tuber, were intensively studied by Goss (1924). The

Figure 1. Potato spindle tuber disease. Symptoms of PSTV in potato (*Solanum tuberosum*). (*a*) Foliage symptoms in cv. Irish Cobbler. Healthy plant at left. (*b*) Tuber symptoms. Upper row: cv. Saco; lower row: cv. Kennebec. Left: healthy; center: infected with the type strain; right: infected with the unmottled curly dwarf strain. Courtesy: M. J. O'Brien, U.S. Department of Agriculture, Beltsville, Md.

author reported that cool weather during the early growth of plants masked the symptoms of spindle tuber so that even with high temperatures later in the season the plants did not have severe symptoms. On the other hand, plants that started growth under high temperature conditions, such as occurred with the late-planted lots, showed severe symptoms throughout the season. Similarly, tuber symptoms of plants grown in the greenhouse at a temperature of 25°C were more severe than at 15°C.

Later, Goss and Peltier (1925) reported that, under controlled conditions in the greenhouse, high soil temperatures increased the severity of the tuber symptoms and that high soil moisture tended to produce the same effect. When mosaic and spindle tuber occurred on the same plant, the severity of the mosaic symptoms decreased, while those of spindle tuber increased, at high temperatures. Some of these findings were confirmed and extended in a later study (Goss, 1930a).

The importance of temperature on symptom expression was confirmed in a study in which individual lots of tubers were divided, with one portion being planted in Maine and the other portion in Florida (Gratz and Schultz, 1931). In Florida, the spindle tuber plants never attained a height greater than from one-third to one-half that of normal plants, while in Maine the diseased plants were about two-thirds to three-fourths as large as those from healthy stock. Otherwise, however, the foliage symptoms were practically the same in both sections.

2.1.4   Field Transmission and Spread

Conclusive evidence for the infectious nature of the spindle tuber disease was first presented by Schultz and Folsom (1923a). These authors noticed that if strains free from this malady were planted near stock with a high percentage of diseased plants, a considerable percentage of spindle tuber resulted among the healthy plants "in a few years."

Experiments in which healthy plants were allowed to grow at various distances from spindle tuber plants indicated that the percentage of spindle tuber decreased as the distance from diseased hills increased, and that a higher percentage of infection of healthy hills occurred as the percentage of spindle tuber increased (Schultz and Folsom, 1923a).

In transmission experiments, both in the greenhouse and in the open field, infection was obtained with tuber and vine grafts, with a "leaf--mutilation" technique,[2] and with aphids (species not indicated) (Schultz and Folsom, 1923a).

On the basis of these results (and because of the demonstrated perpetuation of the disease in the tubers), the authors characterized spindle tuber as a degeneration disease. Later results by other workers demonstrated that spindle tuber was readily transmitted by contact of diseased with healthy

---

[2] Leaf-mutilation inoculation consists of bruising the healthy leaves with the fingers or palms and applying to the spongy mass of mutilated leaf tissue juice that has been expressed from a diseased plant (Schultz and Folsom, 1923a).

plant parts, including foliage (see below). Because this fact was unknown at the time, the results of transmission tests with arthropod vectors must be regarded with some scepticism. In the early tests of Schultz and Folsom (1923a) with aphids, for example, healthy and diseased plants were kept in the same insect cage (Schultz and Folsom, 1925). Thus the possibility of inadvertent transmission of the disease by contact between healthy and infected foliage cannot be ruled out. When these experiments were repeated with aphids being transferred from diseased to healthy plants grown in *separate* cages, results indeed were negative or, if positive, no different from uninoculated control plants (Schultz and Folsom, 1925). In similar tests performed in 1923, however, apparent transmission of spindle tuber by aphids took place in a few cases, but results were scored solely by tuber shape which, in most cases of apparent transmission, was termed as *somewhat spindling*. Thus evidence for aphid transmission remained ambiguous.

The authors also investigated whether the mere contact of seed pieces, roots, and shoots of spindle tuber plants with those of healthy plants was sufficient to produce infection in the absence of aphids. Here contact of freshly cut surfaces was not intimate, as in tuber grafting. No transmissions were obtained (Schultz and Folsom, 1925). Werner (1926) reported that spindle tuber was artificially transmissible by tuber grafts and that in the field the disease was spread from affected plants to healthy plants in adjacent rows and to plants in rows some distance away. The extent of transmission decreased with the distance from the spindle tuber stock. When healthy plants were caged, no transmission occurred from diseased plants even if in close proximity. No transmission tests with aphids were reported, but because aphids were present in very small numbers in the experimental fields, the author considered it "very unlikely that they were responsible for more than a trifle of the transmission" that occurred.

Goss (1926a, b) presented evidence that potato spindle tuber was readily transmitted when the freshly cut surfaces of healthy and infected seed pieces were rubbed together. Transmissions were also obtained by cutting healthy seed with a knife previously used to cut infected tubers. The author suggested that cutting knives and seed-piece contact might prove to be sources of infection of plants in the field.

These results appeared to be in conflict with those obtained earlier in Maine which had indicated that spindle tuber could not be transmitted by cutting knives (Folsom, 1923; Schultz and Folsom, 1925) or by contact with infected seed pieces (see above).

In light of Goss' results, Bonde (1927) reinvestigated in Maine the transmissibility of spindle tuber by cutting knives and seed-piece contact and was able to confirm the results of Goss (1926a, b).

Transmission of spindle tuber by grasshopper (*Melanoplus* spp.) was reported by Goss (1928), and later transmission by flea beetles (*Epitrix cucumeris* and *Systena elongata*), tarnished plant bugs (*Lygus pratensis*), the larvae of the Colorado potato beetles *(Leptinotarsa decemlineata)*, and leaf beetles (*Disonycha triangularis*) (Goss, 1930b, 1931).

Here again, the possibility of mechanical transmission during handling of plants must be considered. In this connection it may be significant that of the seven arthropod species tested, all except one apparently transmitted spindle tuber, and this one species (the leafhopper *Euscelis exitiosus*) was tested on only three plants, whereas the species with which positive results were achieved were tested on 12 to 118 plants (Goss, 1931). On the other hand, introduction of saliva secreted by grasshoppers that had fed on spindle tuber plants into healthy plants did not cause any infections. Presumably, the extent of handling of plants in these tests was at least comparable to that necessary in tests with live insects. Thus if transmission did occur by mechanical means, and not by insect feeding, the tests with saliva should have resulted in a comparable number of transmissions as were observed with live insects.

Clearly then, these early tests do not conclusively demonstrate whether potato spindle tuber is transmitted in nature by an arthropod vector of one kind or another. Surprisingly, no newer experiments on insect transmission of the disease seem to have been reported except for a brief statement by Merriam and Bonde (1954) to the effect that "the disease was not readily transmitted by aphids, grasshoppers, or flea beetles."

Although mechanical transmission of spindle tuber through the foliage had been shown to occur already by Schultz and Folsom (1923) when infective sap was applied to mutilated leaves of healthy plants for some time, this method was not considered to be important under field conditions (Manzer and Merriam, 1961). Goss (1931) demonstrated mechanical transmission by the following methods: (1) needle pricks, (2) leaf mutilation, (3) insertion of crushed leaves into stem, and (4) insertion into stems of cotton saturated with extract from infected plants. However, it was only when Bonde and Merriam (1951) demonstrated that relatively high percentages (33 to 50%) of transmission could be obtained by either rubbing together freshly cut surfaces of healthy and diseased seed pieces [in conformation of the results of Goss (1926a, b)] or by bruising young sprouts of healthy plants and contaminating them with sap from infected tubers that the concept of mechanical transmission as an important means of spindle tuber spread under field conditions gained acceptance.

Even higher percentages of transmission (80 to 100%) were achieved by Merriam and Bonde (1954) by brushing actively growing healthy plants with diseased foliage, further emphasizing the importance of mechanical transmission for the spread of spindle tuber in the field. Furthermore, the authors stated that the disease could be spread by driving contaminated tractor wheels over the foliage of healthy plants, indicating that the tractor-drawn machinery used in producing the crop might be important in the dissemination of spindle tuber (Merriam and Bonde, 1954).

Further detailed studies on the mechanical transmission of the disease under field conditions were reported by Manzer and Merriam (1961). They concluded that the disease could be disseminated very readily in potato fields by contact of healthy vines with contaminated cultivating and hilling equipment. In the Katahdin and Kennebec varieties nearly 100% transmission was recorded in tests that simulated excessive contact of large vines with contaminated equipment, whereas under conditions of less severe vine contact, a lesser amount of transmission was observed.

In summary, the importance of mechanical transmission for the spread of potato spindle tuber disease in the field appears to be well established, whereas the role of arthropod species as vectors of the disease agent is somewhat equivocal.

One study that seriously conflicts with the results of other investigators needs consideration here. MacLachlan (1960), who studied potato spindle tuber in Eastern Canada, concluded that the disease was transmissible by grafting and by dodder, but not by mechanical means or by aphids.

In graft transmission experiments from plants that were showing spindle tuber symptoms to USDA seedling 41956, two types of symptoms were obtained: (1) symptoms resembling those of "purple top or bunchy top, the disease attributed to aster yellows infection," and (2) severe spindling and stunting of plants with marked proliferation of axillary buds, symptoms somewhat resembling "those of plants infected with witches' broom virus." The author concluded that a virus of the yellows type was closely associated with the condition recognized as potato spindle tuber in Eastern Canada. Neither of the two symptom types observed in USDA seedling 41956 resembled those caused by spindle tuber infection in the field, and no symptoms that could be interpreted as being due to spindle tuber infection were observed on the foliage of the cultivars Kennebec, Sebago, Irish Cobbler, or Green Mountain in the greenhouse. Also, tubers harvested from plants arising from spindle tubers were usually of normal shape. In light of these observations it appears most doubtful that the condition(s) studied by Mac-Lachlan (1960) was related to the potato spindle tuber disease described earlier in the U.S.A. or investigated later in Canada (see Chapter 3).

The spindle tuber agent, apparently, cannot be transmitted through the soil. In limited greenhouse tests, Goss (1931) determined that healthy seed planted in soil previously containing spindle tuber or unmottled curly dwarf plants always produced healthy plants.

2.1.5 Natural Hosts

In nature, potato is the only known host of the spindle tuber agent. No weed species that might serve as reservoirs of the agent have been reported.

Extensive trials to find resistance in *Solanum tuberosum* to the potato spindle tuber agent were largely unsuccessful. None of the commercial varieties grown at the time showed any resistance to the disease when exposed to infection (Werner, 1926). Among 2037 seedlings from 24 family lines, 274 seedling selections, and 34 named cultivars of *Solanum tuberosum* that were tested for resistance to mechanical inoculation with the spindle tuber agent, only nine showed a high level of resistance after two inoculations (Manzer *et al.*, 1946). Results of reinoculation tests on selections surviving initial exposure showed, however, that field resistance and not immunity was involved.

2.1.6 Strains

2.1.6.1 Unmottled Curly Dwarf

In addition to the type strain so far considered [as maintained under the name *spindle tuber* in the Schultz Potato Virus Collection (Webb, 1958)], Schultz and Folsom (1923a) described another potato disease with similar, but more severe, symptoms. This condition was named *unmottled curly dwarf* because of its similarity to "curly dwarf" of potato described by Orton (1914) but lacking mottling of leaves. Because this condition remained true to type for three years, Schultz and Folsom (1923a) considered it a single disease, not a combination of diseases. Symptoms consisted of "pronounced dwarfing, spindliness, dark green color of the foliage early in the season, wrinkling, burning, somewhat premature death, and spindling, gnarled, and cracked tubers" (Schultz and Folsom, 1923a). Unmottled curly dwarf was shown to be transmissible by the leaf mutilation method and by aphids (Schultz and Folsom, 1923b). Later, Folsom (1926) pointed out that the symptoms of spindle tuber and unmottled curly dwarf were similar and that the difference was chiefly one of degree, with a few additional related symptoms characterizing unmottled curly dwarf, particularly the appearance of serious lengthwise growth cracks in the tubers (Fig. 1). Goss (1930a) pointed out that the overlapping of symptoms of the two diseases, together with the variation in symptoms occurring in different varieties and in individual plants of the same variety, made it impossible in many instances to make an accurate diagnosis. He stated that it was quite probable that both diseases had been studied under the name of spindle tuber and that unmottled curly dwarf had often been considered as *severe spindle tuber*. Goss (1930a) showed in several thousands of inoculations with both diseases that each maintained its own characteristics upon transmission and that, evidently, unmottled curly dwarf was not the result of a gradual and progressive increase in the severity of spindle tuber. Today, unmottled curly dwarf is generally considered to be a severe strain of potato spindle tuber (Diener and Raymer, 1971).

2.1.6.2 Mild Strains

The existence of mild strains of the potato spindle tuber agent became evident only after tomato was used as a test plant for the pathogen (see Chapter 3). Fernow (1967) reported that use of tomato resulted in the detection of only about 13% of the infected clones tested and attributed the poor results to the existence of a strain that produced symptoms in tomato so mild that they were easily overlooked. The author, after ruling out other explanations for these results, such as the presence of inhibitors of infection in some clones, adopted and subsequently confirmed a working hypothesis that there are two strains (or groups of strains) of the spindle tuber agent in potatoes. One of these, which can be designated as "severe" in tomato, induces extreme shortening of internodes, causing a rosette appearance, severe epinasty and downward curling of leaves, severely wrinkled leaves, and shortening of petioles and midribs. Severe necrosis of stems, petioles, and midribs often occurs, but usually 5 to 7 days after the other symptoms. This type was found only rarely in inoculations from field-grown potatoes.

The other type, designated as "mild" in tomato, induces symptoms in tomato that are slow to develop and are so mild that they are easily overlooked. They consist of slight epinasty and twisting of terminal leaflets and a general reduction in growth. Necrosis does not develop with the mild strain (Fernow, 1967). The author found that cross protection by this mild strain against the severe strain occurred, but that it was usually partial and temporary. The author developed a double-inoculation technique and demonstrated its validity and value for the elimination of spindle tuber from seed stocks before planting (Fernow *et al.*, 1969) (see Chapter 3).

Singh (1970a) and Singh *et al.* (1970) established by use of Fernow's double-inoculation technique that, of field samples found to be infected, 86 to 92% harbored the mild strain, and only 8 to 14% the severe strain. Greenhouse and field tests indicated that the mild strain, so named because of mild symptoms on tomato, was mild in potato as well.

2.1.7 Economic Losses

While the potato spindle tuber disease does not cause total destruction of the crop and does not cause storage losses, it does cause a serious reduction in the total production (Werner, 1926). A further loss to the producer results from the smaller size of the tubers that, together with fewer potatoes per hill, results in a lower yield of marketable size potatoes. "A lot seriously affected with spindle tuber is sold at a discount because of the rough shape and abnormal appearance of the potatoes. The abnormal appearance makes the tubers unattractive as well as a commercially nonstandard article" (Werner, 1926).

The extent of the loss, of course, depends on the incidence of diseased plants in the field. No accurate estimates of losses appear to have been published, however, during the early years of research on the disease. It is clear, nevertheless, that potato spindle tuber was a serious disease in the irrigated potato growing areas of Eastern Nebraska (Goss and

Peltier, 1925) and, prior to 1923, in Aroostock County, Maine (Folsom, 1923), where it was not difficult to find fields that contained 25 to 90% of the plants manifesting symptoms of the disease (Gratz and Schultz, 1931). For 25 years thereafter, the disease was not a serious problem in Maine because of effective control measures instituted in the early 1920s (Bonde and Merriam, 1951). In the early 1950s, however, the disease again became increasingly prevalent in Maine, appeared in a number of farmers' seed stocks, and was found to be quite prevalent in some seed stocks of the newly released Kennebec variety (Bonde and Merriam, 1951). An estimate of 2.6% reduction of yield was obtained by LeClerg et al. (1944) on the basis of a 4% incidence of infected plants of the varieties Irish Cobbler, Green Mountain, Katahdin, Chippewa, and Triumph in Alabama, Florida, Louisiana, Maine, Maryland, New Jersey, and Virginia. When the incidence of infection was 100%, yield reduction varied from 20.3% for Katahdin in Florida to 68.5% for Triumph in Louisiana (LeClerg et al., 1944).

According to Hunter and Rich (1964a), most of the research done with spindle tuber before 1960 had been done with plants that were infected with one or more other pathogens, notably potato virus X and possibly potato viruses S, M, and others. To determine the effects of spindle tuber on growth and tuber yield in the absence of other viruses, these workers used the Saco variety because of its apparent immunity to mild and latent mosaics (viruses A and X) (Akeley et al., 1955) and to virus S (Bagnall et al., 1956). They found that infection with potato spindle tuber caused a marked reduction in yield (average 64.5%) that was due in part to fewer tubers per hill and in part to reduction in size of tubers.

More recently, Singh et al. (1971) reported that in field trials, three isolates of the mild strain of the spindle tuber agent reduced yield in the variety Saco by 17, 24, and 24%, respectively, and the severe strain reduced yield by 64%. The authors estimated an incidence of 3.8% of potato spindle tuber disease among the three major varieties Kennebec (3.3%), Katahdin (2.5%), and Netted Gem (Russet Burbank) (4.6%). On the basis of these estimates, the ratio of mild to severe strains observed, and the measured yield reduction with each strain (see above), Singh et al. (1971) calculated a hypothetical loss of about 1% of the potato crop.

These studies emphasize the importance of control of spindle tuber. As long as control is effective in keeping the incidence of the disease low, yield losses are of little consequence. Inadequate control, however, can lead to catastrophic losses in a relatively short time.

2.1.8 Control Measures

With the realization that potato spindle tuber is an infectious disease, certain control measures suggested themselves.

Schultz and Folsom (1923a) had already conceived and tested a number of these, and it is interesting to note that even today successful control of potato spindle tuber disease is based on principles very similar to those elaborated by Schultz and Folsom in the 1920s. The authors demonstrated that is was possible to eliminate a large percentage of the spindle-shaped tubers from seed stocks and so reduce the percentage of diseased plants. They noted, however, that tuber selection alone did not usually result in eliminating the disease because tuber symptoms were not always conspicuous and more importantly, because an unknown number of the normal-shaped tubers might in fact be infected due to late season transmission in the field (which does not result in current season tuber symptoms).

Folsom (1923) stated that there was no way of learning ahead of time how large a proportion would be diseased in the progeny of apparently healthy tubers from a field containing diseased plants and that as little as 5% of the hills diseased one year had contaminated as high as 70% of the next year's progeny of good-shaped tubers.

More successful in controlling the disease was a program in which healthy-appearing hills, which were not next to diseased hills, were selected. Here, success depended largely on the extent of disease occurrence. In a field of one variety when less than 1% of the hills were diseased, it was possible to select actually healthy hills without any failure, but in a field of another variety, the presence of 5% made it impossible by hill selection to avoid a considerable increase of the disease in the stock (Folsom, 1923).

Other means of control that were developed early consisted of prompt removal, or roguing, of diseased hills and of the growing of selected healthy stock in isolation.

Adherence to these control measures in Maine and elsewhere apparently was effective in eliminating the spindle tubr disease as a serious threat to the potato grower (Bonde and Merriam, 1951).

Control measures used today are still based on the principles developed by the early workers, namely "selection of a strain entirely (or practically) free from spindle tuber and then growing it in an isolated and severely rogued seed plot" (Werner, 1926), except that, in some cases, indexing of seed stocks on tomato or even analysis by gel electrophoresis (see Chapter 3) is used, instead of visual examinations, to eliminate infected plants from next year's seed stock.

2.2   TOMATO BUNCHY TOP DISEASE

2.2.1   Geographic Distribution

The tomato bunchy top disease is known to occur naturally only in South Africa. It was first described by McClean (1931), who reported that the disease had been recorded for the first time from the Airlie district of the Eastern Transvaal during the early part of 1926 and that it had been found to be more or less generally distributed through the low-lying districts of the eastern part of the Transvaal. At that time (McClean, 1931), no records existed of its occurrence outside this area, either in the same province or other parts of South Africa. Later, the disease had been recorded also from the coastal area of Natal (McClean, 1948). No more recent reports appear to have been published on this disease.

2.2.2   Symptomatology

The general appearance in the field of tomato plants infected with bunchy top disease depends on the stage and duration of infection (McClean, 1931). Characteristic symptoms are severe stunting of the entire plant, in particular the production of small leaves, a necrosis of the leaves and stems, and various forms of leaflet distortion, such as curling and an abnormal unevenness of the surface (McClean, 1931). Chlorosis or mottling are not features of the disease.

The first indication of the disease is a sudden and almost total cessation of growth at the branch extremities, with the result that at these points the leaves become closely crowded, giving to the plant a bunched appearance which is typical of the early stages of infection. Leaves fully developed at the time of infection do not undergo any change but remain normal and healthy in appearance. In the condensed region, however, there is a progressive decrease in leaf size, and closer crowding of the leaflets on the rhachis. The leaflet margins become crowded toward the undersurface, the tips frequently are twisted downward, and the surfaces show a puckered condition. Axillary buds, particularly those of the lower leaves, are forced into early activity (McClean, 1931).

The check on the upward growth of the axis is not permanent and is followed by a definite elongation of the internodes, with the production of a somewhat spindling type of growth. A plant at this stage shows a lower region that is apparently normal, a middle region with condensed axis, bunching of foliage, a progressive reduction in leaf size, and various forms of leaf distortion, and an upper region in which the internodes have again lengthened, and which is characterized by a thin axis and small leaves that show little or no distortion.

As the diseased plant matures, the older leaves, including apparently healthy ones and those in the bunched region on the main axis, die off and are shed. Thus a plant in an advanced stage of the disease bears only the typical dwarfed leaves produced after infection.

Another characteristic symptom of the disease is necrosis of the leaves and stems of plants. Very narrow brown or black streaks appear along the undersurface of the rhachis and leaflet veins some days following the first signs of infection. This is accompanied or followed by the development of broad black streaks along both surfaces of the rhachis and by the blackening of leaflet veins. Both main and lateral veins may be affected. Necrosis is characteristic of the early stages of the disease; growth produced in the elongation phase usually shows no necrosis (McClean, 1931).

Diseased plants produce apparently normal flowers and the setting of fruit is not inhibited. Fruits are small, sometimes distorted, and are either seedless or else contain a few small seeds, only a portion of which may be fertile.

Based on field observations and greenhouse studies, McClean (1931) concluded that symptoms were most pronounced in plants grown under optimal conditions. Thus symptom expression was favored by high light intensity, relatively high temperature, and good soil fertility.

2.2.3 Field Transmission and Spread

Field observations showed that the bunchy top disease occurred over a wide range of soils and that, in support of experimental studies, the disease was not soil borne (McClean, 1931). Usually, there was no evidence of the disease until 5 weeks after transplanting when the plants came into flower. Judged from their size and the period elapsed from transplanting, it appeared likely that plants became infected at the time of planting out or subsequent to it (McClean, 1931).

Because tomatoes are almost in a continuous state of cultivation in the area, it is probable that the diseased plants in an earlier crop are often the source from which infection starts. It is apparent, however, that the tomato is not the only reservoir of infection, for bunchy top may appear even in the absence of earlier plantings (McClean, 1931). The author proposed that the disease might be carried in solanaceous weeds or in some indigenous species.

In greenhouse experiments, McClean (1931) demonstrated that the disease was readily transmitted by grafting or by direct juice inoculation, but not by the seed, soil, nematodes, or certain aphids. These results, as well as field observations, suggest that in certain instances the human factor has been important. This applies particularly to the few serious outbreaks, whose localized nature and comparatively rare occurrence suggest some accidental cause, such as seed bed infection, or the contamination of seedlings during transplanting or subsequently in cultivation operations. The ease with which the disease is transmitted by mechanical means makes this a likely possibility. On the other hand, the initial outbreaks which involved only occasional diseased plants widely scattered in the field suggested to McClean (1931) rather the influence of insects.

2.2.4 Natural Hosts

The agent of the bunchy top disease can be transmitted experimentally to a number of plant species in addition to tomato (see Chapter 3), but it is not known whether any of these or other plant species harbor the agent in nature (McClean, 1931, 1935, 1948).

2.2.5 Strains

No definite evidence exists as to whether different strains of the bunchy top agent exist. The only suggestion in McClean's experiments for the possible existence of a severe strain occurred in experiments with *Physalis peruviana*, a solanaceous plant species in which the bunchy top agent replicates but causes no symptoms or at most rather indistinct ones (McClean, 1935). On three occasions, untreated plants of this species "spontaneously" developed a condition almost identical to that produced in tomato by the bunchy top agent. In all cases, transfers from plants with this condition ("severe disease") to tomato indicated that these plants had acquired the bunchy top agent, yet transmission from plants with "severe disease" to healthy *P. peruviana* did not necessarily produce the symptoms of severe disease. McClean (1935) offered two explanations to account for "severe disease": (a) that it represented a reaction to some modified form of bunchy top agent, and (b) that it represented the multiple effect of normal bunchy top agent and a second virus. The author, however, considered the second alternative unlikely because the chance meeting of the two postulated pathogens occurred only in three untreated plants and never in the many plants that were artificially infected with the ordinary strain of bunchy top. On the other hand, this observation could be explained if "severe disease" were caused by a modified form of the bunchy top agent because, in this case, the presence of the less virulent strain in the plants might protect them from invasion of the more severe strain (McClean, 1948).

2.2.6 Economic Losses

According to McClean (1931), serious outbreaks of the disease occurred infrequently and, in addition, were extremely localized, affecting one or more fields on individual estates, while only occasional diseased plants had been evident in fields on adjacent holdings. The author pointed out the interesting contrast with insect-borne virus diseases in South Africa, such as peanut rosette or streak disease of maize and sugar cane which, when outbreaks had attained epiphytotic proportions, had been almost invariably general over large districts.

   In most cases, however, the incidence of the disease was exceedingly low; plants with primary infections were widely scattered in the field. Typically, little increase in the number of plants with symptoms occurred until after the plants were past full bearing, and then an appreciable increase in the incidence was often noticeable (McClean, 1931). No newer data on the incidence of the bunchy top disease and on economic losses suffered by growers are apparently available.

2.2.7 Control Measures

The infectious nature of the disease and its comparatively ready transmission by direct mechanical means stress the importance of avoiding successive handling of diseased and healthy plants, particularly in handling seedlings while in the seedbed or subsequently in transplanting (McClean, 1931).

Field observations indicated that the first diseased plants that developed became infected after transplanting. If so, destruction of such sources of the disease as exist in old fields should lead to a reduction in the number of early cases of bunchy top (McClean, 1931).

No information seems to exist as to whether the control measures recommended by McClean were adopted and, if so, whether they were successful in controlling outbreaks of the disease.

2.3 CITRUS EXOCORTIS DISEASE

2.3.1 Historical

Apparently, the first report on exocortis disease appeared in 1948. Fawcett and Klotz (1948) described a condition of bark shelling of trifoliate orange [*Poncirus trifoliata* (Linn.) Raf.] that had been noted "occasionally for many years in California." The authors stated that this condition was known as exocortis [from exo = outside, and corti(ci)s = pertaining to the bark] and that the nature or cause of the condition was not yet known, aside from the assumption that it was due either to a "genetic factor or to a virus."

Although exocortis was not described until 1948, its existence as far back as the early 1920s in California and South Africa can be deduced from the literature (Knorr and Reitz, 1959).

In 1949, a detailed study appeared by Benton *et al.* (1949, 1950) on "stunting and scaly butt of citrus associated with *Poncirus trifoliata* rootstock" in New South Wales, in which the authors presented evidence that this condition was transmissible and therefore "virus" caused. The authors considered scaly butt to be identical with exocortis as described in California.

A disorder in Brazil affecting citrus trees budded to Rangpur lime rootstock, with symptoms strikingly similar to those of exocortis, was described by Moreira (1955). On the basis of this similarity, as well as of the observation that the same top varieties caused the disease on the two rootstocks (*P. trifoliata* and Rangpur lime), Moreira (1955) concluded that the disease in Brazil was caused by the exocortis "virus."

Olson and Shull (1956) described a condition of citrus trees in Texas, that had been budded to Rangpur mandarin-lime rootstock; and these symptoms ("bark shelling") closely resembled those described by Moreira (1955). Although Olson and Shull (1956) accepted Moreira's conclusion that the disorders in Brazil and Texas were identical and due to exocortis "virus," the senior author stated that in his opinion more definite proof of this identity was necessary. Similar reservations were voiced by Childs *et al.* (1958), who stated that the exocortis-type scaling exhibited by certain citranges (Bitters, 1952; Olson and Shull, 1956), lemons (Benton *et al.*, 1949, 1950), and limes (Olson and Shull, 1956) "may or may not be caused by the exocortis virus, a point not established."

Further circumstantial evidence for the identity of the exocortis agent with that causing the Rangpur lime disease was reported by Reitz and Knorr (1957), and the belief that the two diseases were caused by the same pathogen became widely accepted (Calavan and Weathers, 1961). Later it was shown that the exocortis agent affected a number of other citrus species and varieties (see Chapter 3), in addition to trifoliate orange and mandarin-lime rootstocks.

2.3.2 Geographic Distribution

Exocortis disease has been reported to occur in all major citrus-growing areas of the world, such as Australia (Benton *et al.*, 1949, 1950), Brazil (Moreira, 1955), Argentina (Knorr *et al.*, 1951), California (Fawcett and Klotz, 1948), Texas (Olson and Shull, 1956), Florida (Reitz and Knorr, 1957), South Africa (McClean, 1950), Corsica (Vogel *et al.*, 1965), Spain (Planes *et al.*, 1968), Japan (Yamada and Tanaka, 1972), and Taiwan (Ling, 1972), among others.

2.3.3 Symptomatology

In many citrus species and varieties, infection with the exocortis agent does not result in obvious macroscopic symptoms (Olson, 1968), and its presence in these symptomless carriers may be determined only by inoculating tissue or extracts of the suspected host into sensitive indicator plants, a process that is generally called *indexing*.

In essence, this is what Benton *et al.* (1949, 1950) did when they grafted Washington Navel orange (the symptomless carrier) onto the sensitive *P. trifoliata* rootstock, and thereby demonstrated that scaly butt (or exocortis) was a transmissible disease.

In such a combination, symptoms of the disease consist of scaling of the bark below the graft union and stunting of the tree (Fig. 2). The first sign of scaling is the appearance of small, dead areas in the outer bark of the stock below the bud union (Benton *et al.*, 1949). These are oblong, elongated, or scalelike. Subsequently, longitudinal fissures connected by horizontal cracks develop around the margins of the dead areas, separating them from the adjacent bark, and the scales so formed partially lift away from the living bark beneath. The process of scale formation goes on intermittently, new scales being cut off beneath the old rather irregularly. In the common severe type of exocortis, the first scales are formed at about ground level and the condition spreads almost at once upward to the bud union. In other cases, scaling may start at or near the bud union or it may start at one side and remain restricted for a number of years, only slowly extending up to the bud union or around the whole circumference (Benton *et al.*, 1949). Not infrequently whole roots die out. In trees that develop scaling, the diameter of the stock is the same as or only slightly greater than that of the scion, which is in pronounced contrast with the appearance of trees in which exocortis-free scions are budded onto trifoliate orange rootstocks. In such trees the rootstock outgrows the scion and in mature trees is several times greater than the scion in diameter and is deeply ribbed (Benton *et al.*, 1949).

Stunting of the scion usually becomes apparent at the time that scaling is first seen. In the most extreme cases the trees do not exceed 4 to 5 feet in height and are rather sparse of foliage. Stunted trees crop well for their size, and the fruit is of normal size and of particularly high quality (Benton *et al.*, 1949).

Budding of exocortis-infected scions to Rangpur lime rootstock results in symptoms very much like those observed in trees on *P. trifoliata* rootstock. Trees become stunted and show gumming and bark shelling of the butt below the bud union; yellowing of foliage and off-season flowering may occur in some strongly affected trees; only in very severe cases do trees die (Moreira, 1955).

Moreira (1961) demonstrated that topworking of exocortis-infected trees with Rangpur lime or trifoliate orange resulted in yellow blotches on the bark of the topworked varieties and that later their bark split and shelling appeared.

Figure 2. Citrus exocortis disease. Symptoms of CEV in Valencia orange on trifoliate orange rootstock. (*a*) Healthy control tree; note normal "benching" of rootstock. (*b*) CEV-infected tree; note smaller diameter of trunk and scaling of bark. Courtesy: S. M. Garnsey, U.S. Department of Agriculture, Orlando, Fla.

Later it became evident that the natural host range of the exocortis agent extended to many additional citrus scion-rootstock combinations (see Chapter 3). In all of these, symptoms are essentially as described for trees on trifoliate orange or mandarin-lime stock, except that the severity of symptoms may vary greatly and, in some cases, stunting may occur without much bark scaling or, vice versa, scaling without severe stunting.

As shown by Weathers *et al.* (1965), symptoms in any given scion-rootstock combination are modified depending on the nutritional status of the host tree. In trees grown on trifoliate orange rootstock, for example, excess nitrogen and phosphorus favors symptom formation (Weathers *et al.*, 1965).

2.3.4   Field Transmission and Spread

For some time it was believed that exocortis did not spread in the field. Thus Benton *et al.* (1950) considered the lack of field spread as an indication against the theory that exocortis was caused by a "virus." In one of their experimental plantings, however, 9% of trees budded with material from exocortis-free source trees developed exocortis symptoms (Benton *et al.*, 1949). In discussing these results, the authors suggested that it was highly probable in this experiment that two lots of trees could have been mixed at planting (Benton *et al.*, 1950).

Some observations that suggested the possibility of field spread were reported by Calavan *et al.* (1959). They noted that in a nursery some of the trees grown near to and downhill from exocortis-infected trees became infected, while none of the trees grown uphill from diseased trees developed exocortis. The authors suggested on the basis of this observed unidirectional spread of the disease within the nursery that exocortis transmission might not have depended on root grafts alone but might have been accomplished by a vector which moved principally in the direction of water flow.

Moreira (1955), however, considered it unlikely that the exocortis "virus" was transmitted by insects because in Brazil many trees of susceptible combinations had grown for 20 years beside affected trees and were still free of exocortis (Moreira, 1959). In Australia, also, observations indicated that field spread of exocortis did not readily occur (Fraser and Levitt, 1959). Only one case was located during 15 years where exocortis had developed in a tree over 10 years old, previously known to be free of the disease. But even in this one case the possibility of infection by root grafting with adjacent affected trees could not be ruled out. Consequently, at that time, it was generally believed that exocortis was distributed principally, if not entirely, by the use of infected propagative material and by natural root grafts (Calavan and Weathers, 1959).

In 1967, however, Garnsey and Jones (1967) observed natural infection of citron (*Citrus medica*) plants under greenhouse conditions and demonstrated that neither seed transmission nor an insect vector was involved. The authors discovered that the "virus" was mechanically transmitted as a contaminant on budding knives, even though it was not transmitted by conventional sap inoculation of leaves.

These results strongly suggested the possibility that exocortis might be spread in commercial operations by contaminated tools.

2.3.5 Natural Hosts

Although the exocortis agent is widespread in many tolerant varieties, its presence becomes evident only when such symptomless carriers are grafted onto sensitive rootstocks. Thus in the field the exocortis disease appears as a typical graft union problem.

Sensitive rootstocks, aside from trifoliate orange, are various citranges and mandarin limes (Knorr and Reitz, 1959), certain varieties of sweet lime, sweet lemon, and Cuban shaddock (Weathers and Calavan, 1961). In some rootstocks, considered to be tolerant to exocortis, subtle symptoms nevertheless may occur. Measurable stunting occurred, for example, when exocortis-infected lemon was grafted onto sweet orange, grapefruit, or sour orange rootstock (Calavan and Weathers, 1959). Similarly, exocortis-infected Washington navel orange trees on Cleopatra mandarin or sweet orange rootstocks showed some stunting but no bark scaling, and infected trees on sweet orange rootstock bore less fruit than did uninfected trees (Sinclair and Brown, 1960; Olson, 1968).

2.3.6 Strains

Based on field observations, Fraser and Levitt (1959) suggested that exocortis existed in a number of strains with the following variations:

1. Development fairly rapid (2 to 4 years from time of budding) and preceded by stunting. Scaling starting at the bud union and rapidly involving all of the butt to soil level and a little below. Scion and stock diameters much the same.

2. Development slower (6 to 8 years) but of the same general type.

3. Development slow (6 to 8 years). Scaling starting at a point below the bud union and often involving only a relatively small area at first. The stock in this case is moderately overgrown.

Another variable is tree size. Small, medium, and fairly large trees are associated with the different types of scaling.

Weathers and Calavan (1961) distinguished strains differing in incubation periods in susceptible hosts and in their reactions to phloroglucinol-HCl in *P. trifoliata*. Calavan and Weathers (1961) noted that the time required for symptoms to develop, as well as the severity of these symptoms, was consistent for each exocortis source. They concluded that different sources carried different strains of exocortis "virus." A particularly virulent strain, carried by old-line Eureka lemon trees, had a shorter incubation period in most exocortis indicators, 2 years or less, than the strains carried by many other sources.

Later, when the Etrog citron test (Fig. 3) became available (see Chapter 3), strains of exocortis "virus" were detected that did not cause any detectable symptoms in trees on *P. trifoliata* or Rangpur lime rootstocks (Calavan *et al.*, 1964).

Further evidence for the existence of numerous strains of exocortis "virus" was presented by Salibe and Moreira (1965a). These authors also reported that buds carrying mild strains could be obtained from trees infected with a severe strain, suggesting that mild and severe strains might exist in the same tree. Occasionally, apparently healthy buds could also be obtained from trees infected with exocortis "virus," indicating that the agent might, in some instances, not be completely distributed through infected trees.

2.3.7 Economic Losses

Fawcett and Klotz (1948), in their original report on exocortis disease, stated that in the comparatively few citrus orchards in which trifoliate orange rootstock had been located in southern California, possibly 5 to

Figure 3. Citrus exocortis disease. Healthy (left) and CEV-infected Etrog citron (*Citrus medica*); note epinasty and stunting (typical reactions to severe strain). Courtesy: S. M. Garnsey, U.S. Department of Agriculture, Orlando, Fla.

25% of the trees had exocortis. These were much smaller trees and usually showed lack of vigor.

In a testing program in Florida, of 620 mature trees of 27 different citrus varieties, 55% were shown to be infected with exocortis "virus" (Norman, 1965).

Calavan et al. (1968) studied the effect of exocortis on production and growth of Valencia orange trees on trifoliate orange rootstock. Their results indicated that exocortis reduced both growth and fruit production, at least during the first decade, that exocortis caused larger fruit size than in the controls in some years, and that the tendency to alternate bearing, noted in the controls, was absent or less prevalent in exocortis-infected trees. Under the conditions of the experiment, the effect of exocortis infection was greater on tree growth than on fruit yield. This relationship suggests that exocortis reduced production by stunting trees and reducing the fruit-bearing surface (Calavan et al., 1968).

No estimates of overall crop losses due to exocortis infection appear to have been published.

2.3.8 Control Measures

After their recognition of the infectious nature of exocortis disease, Benton et al. (1950) recommended the use of buds from unaffected trees on trifoliate orange stock as a means of obviating exocortis in oranges, grapefruit, and mandarins.

With the development of indexing procedures (see Chapter 3), faster and more reliable means of detecting exocortis infection of propagating material became available, and application of these procedures undoubtedly helped in reducing the incidence of exocortis disease in orchard trees.

It was not until the recognition that the exocortis agent was mechanically transmissible on contaminated propagating tools (Garnsey and Jones, 1967) that entirely effective control measures could be devised. Roistacher et al. (1969) discovered that mechanical transmission by knife-cuts was prevented by a 1-second dip of the contaminated blades in diluted household bleach or a solution of 2% sodium hydroxide plus 2% formalin.

2.4 CHRYSANTHEMUM STUNT DISEASE

2.4.1 Historical

The first report of chrysanthemum stunt was published by Dimock (1947), who described disease symptoms and stated that the disease, which was of unknown causation, had been recognized since 1945. According to Brierley and Smith (1949), the disease became generally prevalent in greenhouse chrysanthemums in the U.S.A. and Canada in 1946. In 1949, Brierley and Smith (1949) presented evidence that the disease was an infectious one and that transmission by both grafting and juice inoculation was possible.

2.4.2 Geographic Distribution

Chrysanthemum stunt disease is widely distributed in the U.S.A. and is present in Canada (Welsh, 1948).[3] It has been recorded in the Netherlands (Noordam, 1952) and is fairly abundant in England (Smith, 1972).

2.4.3 Symptomatology

Symptoms of the disease (Fig. 4) vary with the variety of chrysanthemum (*Chrysanthemum morifolium*), but usually consist of the following (Dimock, 1947): (1) the young foliage may be paler than normal and has a tendency to more upright growth rather than growing at a wide angle with the stem, (2) diseased plants show stunting in growth after they have been in the soil a few weeks, and at maturity they may in some cases be less than half as tall as normal, (3) buds may form and blossoms open a week or 10 days ahead of those on healthy plants, (4) with varieties possessing red pigment, the red component of the color is badly bleached, (5) with some varieties the blossoms may be greatly reduced in size.

In addition, leaves usually are smaller (Brierley and Smith, 1949), and in the cultivar Mistletoe (and some other cultivars) stunt is accompanied by numerous conspicuous white leaf spots about 3 mm in diameter (Keller, 1951). Some varieties show a marked tendency for

---

[3] Welsh (1948) described a presumably distinct disease of chrysanthemum that he called *stunt-mottle*. Later work (Keller, 1953) indicated that this disease, most likely, was identical with chrysanthemum stunt.

Figure 4. Chrysanthemum stunt disease. (*a*) Healthy (left) and CSV-infected leaves of *Chrysanthemum morifolium*, cv. Mistletoe. (*b*) Healthy (left) and CSV-infected *Senecio cruentus* (cineraria) plants of same age; note severe stunting of infected plant. Courtesy: R. H. Lawson, U.S. Department of Agriculture, Beltsville, Md.

35

axillary bud proliferation and an increase in the number of stolons produced (Keller, 1953).

The disease of chrysanthemum in Canada, called *stunt mottle* (Welsh, 1948), was believed to be distinct from stunt because leaves of all affected varieties "display a mottling that roughly borders the veins of the leaves and that may become a distinct vein-clearing." Possibly the distinction was made because, at the time, leaf spots were not yet recognized as part of the stunt syndrome.

2.4.4 Field Transmission and Spread

Brierley and Smith (1949) clearly showed that the disease was transmitted with cuttings taken from diseased parent stock, but that, in addition, spread ocurred from diseased plants to others not previously affected. These observations suggested that a vector might be involved in the spread of the disease, and, indeed, experiments with aphids seemed to give evidence of transmission by at least one aphid species (*Rhopalosiphum rufomaculatum*) (Brierley and Smith, 1949). The authors' later studies showed, however, that no aphid species transmitted stunt and that the earlier false readings had been due to contamination (Brierley and Smith, 1951).

Because the stunt agent is readily transmissible by juice inoculation, transmission by workers and their tools must be considered. Olson (1949) and Brierley and Smith (1951) showed that the stunt agent was disseminated locally chiefly as a result of cultural operations, such as contaminated knives, tools, and hands.

The stunt agent, apparently, is not seed transmitted (Olson, 1949; Brierley and Smith, 1951).

2.4.5 Natural Hosts

The disease has been described to occur naturally only on various cultivars of chrysanthemum grown in greenhouses and clothhouses. Whether the agent exists naturally in any of the experimental host species (see Chapter 3) is apparently not known.

2.4.6 Strains

No strains deviating from the type strain appear to have been described.

2.4.7 Economic Losses

Only one year after the first observation of the disease it became obvious in several areas in the U.S.A. that the trouble was actually increasing. One year later (1947) the increase in prevalence of the disease was so great as to cause considerable alarm; many cases were noted in which 30 to 60% of the plants were so seriously affected as to be almost worthless (Dimock, 1947). A disastrous epidemic in cultivated chrysanthemums developed during 1945–1947 in the U.S.A. (Hollings and Stone, 1970). Later, production of tested clones greatly reduced the incidence of stunt disease, but it is still of considerable economic importance.

Keller (1953) noted that the rapid dissemination of chrysanthemum stunt was possible because the propagation and distribution phase of chrysanthemum production had become highly centralized in the U.S.A., so much so that one firm was supplying a majority of the chrysanthemum cuttings distributed annually. Evidently, once established in the propagation stock of such a firm, there is every possibility that the pathogen may be distributed to all parts of the country within a short time, thus bringing about a disease of epidemic proportions. Such was the case with chrysanthemum stunt disease in the 1940s.

2.4.8 Control Measures

Dimock (1947), after recognizing the infectious nature of the disease, recommended that the best possible insect control be maintained at all times, that all diseased plants be rogued out, and that stock plants for propagation be retained throughout the year in greenhouses that could be maintained almost completely free of insects.

With the realization that the disease was very contagious and readily transmitted by contact of plants with contaminated hands and tools, pertinent precautions were adopted. These included the use of paper shields, flaming of tools, and use of sterilized soil and propagating media (Keller, 1953).

The most reliable method for stunt control, however, consists of the maintenance, in isolation, of indexed mother blocks as the stock plants, coupled with extreme caution to prevent contamination.

Initially, indexing was performed by grafts between stock and indicator plants of the variety Blazing Gold (Brierley and Smith, 1951).

Later, the variety Mistletoe was preferred because of the distinct leaf-spots occurring in this variety as a consequence of infection with the stunt agent (Keller, 1953).

The rapidity with which these control measures were adopted by commercial chrysanthemum growers undoubtedly was an important factor in vastly improving the health of chrysanthemum stocks (Hollings and Stone, 1973) and in overcoming the threat posed by the stunt disease.

2.5 CHRYSANTHEMUM CHLOROTIC MOTTLE DISEASE

2.5.1 Historical

The chlorotic mottle disease of chrysanthemum was first described as a "new" disease by Dimock and Geissinger (1969), who stated that the disease was first seen in 1967. In a later publication, Dimock *et al.* (1971) reported that the disease had been observed in a number of glasshouses in the southern U.S.A. but only occasionally in the north, and that apparently most, if not all, diseased material originated from clonal stock plants grown in Florida. In New York, the only malady occurred in 1967 in a commercial glasshouse range. Because graft transmission was successful and because no pathogenic bacterium or fungus was found associated with the disease, the authors concluded that the disease had a viral etiology (Dimock *et al.*, 1971).

2.5.2 Geographic Distribution

Apparently, the chlorotic mottle disease of chrysanthemum has been reported to occur only in the U.S.A.

2.5.3 Symptomatology

Affected chrysanthemum plants, cultivar Yellow Delaware, show symptoms that include: (1) mild mottling or variegation of the young leaves and occasional distinct chlorotic spotting, (2) general chlorosis of new leaves, often following mottling of early-formed leaves, (3) dwarfing of leaves, flowers, and the entire plant, and (4) delay in development of blossoms (Dimock *et al.*, 1971).

Expression of symptoms within a group of plants varies at any given time and from time to time. With some cultivars, plants might at one time be very chlorotic, 1 or 2 weeks later appear almost wholly recovered, and then again exhibit severe symptoms (Dimock *et al.*, 1971). Often a parent plant and rooted cuttings appear perfectly healthy and severe symptoms appear only after the cuttings are planted.

The most distinct and reliable foliar symptoms of the disease were observed in the cultivar Deep Ridge (Fig. 5), but symptom development is dependent on appropriate temperature and light conditions (Horst, 1975). Symptoms are best expressed with minimal light intensities of 10,760 lx, photoperiods of 12 hours, and temperatures of 24 to 27°C (Horst, 1975). No symptoms are expressed at constant temperatures less

Figure 5. Chrysanthemum chlorotic mottle disease. Healthy (left) and ChCMV-infected leaves of *Chrysanthemum morifolium* cv. Deep Ridge. Courtesy: R. K. Horst, Cornell University, Ithaca, N.Y.

than 21C and light intensities less than 5,380 lx. Under optimal light and temperature conditions, symptoms develop after 14 days (Horst, 1975).

Other chrysanthemum cultivars show only mild symptoms and still others none at all (Dimock and Geissinger, 1969). Those with no symptoms were infected as demonstrated by grafting to sensitive cultivars (Dimock et al., 1971).

2.5.4 Field Transmission and Spread

Experimentally, the chrysanthemum chlorotic mottle agent can be transmitted readily by grafting (Dimock and Geissinger, 1969), but only with difficulty by sap expressed from diseased plants (Horst and Romaine, 1975). Presumably, therefore, most spread of the disease occurs through infected propagating material. Transmission by hands, tools, or arthropod vectors has not been reported; neither is it known whether the agent is seed transmissible.

2.5.5 Natural Hosts

The disease has been reported to occur naturally only in a few florists' chrysanthemum cultivars, primarily Yellow Delaware and its "parent" cultivar Delaware (Dimock et al., 1971), but all cultivars of florists' chrysanthemum tested were shown to be susceptible (Horst, 1975).

2.5.6 Strains

Horst (1975) suspected that more than one strain of chrysanthemum chlorotic mottle agent existed because, occasionally, inoculated plants exhibited sparse and indistinct symptoms under optimal conditions. More strikingly, some clones of the cultivar Deep Ridge failed to express symptoms even under optimal environmental conditions when inoculated mechanically with chrysanthemum chlorotic mottle agent or with tissue implants from infected chrysanthemums (Horst, 1975). Further experiments demonstrated that this protection was due to a latent infectious agent present in certain clones of several chrysanthemum cultivars. Because with conventional plant viruses cross protection usually is effective only among strains of a single virus and not between different virus species, Horst (1975) suggested that the protective agent was a latent strain of the chrysanthemum chlorotic mottle agent.

2.5.7 Economic Losses

Although the chlorotic mottle disease poses a potentially serious threat to the chrysanthemum grower, vigorous control measures by specialist propagators have largely reduced losses from this disease in commercial chrysanthemum cut- and pot-flower production (Horst *et al.*, 1977). Undoubtedly, the experience gained earlier in controlling the chrysanthemum stunt disease (see above), which nearly destroyed the commercial chrysanthemum industry (Horst *et al.*, 1977), was most useful in rapidly reducing losses due to infection with the chlorotic mottle agent.

2.5.8 Control Measures

As with chrysanthemum stunt, control of chlorotic mottle consists chiefly of using indexing procedures for obtaining a nucleus of disease-free propagating stock (Brierley and Olson, 1956; Horst *et al.*, 1977).

2.6 CUCUMBER PALE FRUIT DISEASE

2.6.1 Historical

A disease in cucumber characterized particularly by a light green color of the fruits, but also with affected flowers and young leaves, was observed in 1963 in two glasshouses in the western part of the Netherlands. Since that time the disease, now called *pale fruit disease*, has been observed in different places over the whole country (Van Dorst and Peters, 1974).

2.6.2 Geographic Distribution

So far the disease has been reported to occur naturally only in the Netherlands (Van Dorst and Peters, 1974).

2.6.3 Symptomatology

In its only reported natural host, cucumber (*Cucumis sativus* L.), the most distinctive symptom is found on the fruits (Fig. 6d). They are pale green in color, retarded in growth, and most are slightly pear

Figure 6. Cucumber pale fruit disease. (*a*) Symptoms at the tip of a CPFV-infected *Cucumis sativus* (cucumber) cv. Sporu plant; (*b*) Tip of a healthy cucumber plant; (*c*) Flowers of a healthy (top) and a CPFV-infected (center and bottom) cucumber plant; (*d*) Fruits of a healthy (left) and CPFV-infected (three fruits at right) cucumber plants. Courtesy: D. Peters, Agricultural University, Wageningen, The Netherlands, and H. J. M. van Dorst, Glasshouse Crops Research and Experiment Station, Naaldwijk, The Netherlands.

42

shaped (Van Dorst and Peters, 1974). Both male and female flowers often are stunted and crumpled (Fig. 6c). The edge of the petals is slightly notched. Developing leaves may be smaller, blue-green, and rugose. The leaf blades are undulated, their edges turned downward, and the tips bent downward or even turned backward. (Fig. 6a). On aging, the leaf symptoms fade and a chlorosis appears. Because the internodes of the younger parts of affected plants are shorter than those of healthy plants, the former often are somewhat stunted. Symptoms develop earlier and are more intense when plants are grown at 30°C than at lower temperatures (Van Dorst and Peters, 1974).

2.6.4 Field Transmission and Spread

Pale fruit disease has been reported to occur naturally only in glasshouses. The disease agent can be transmitted during pruning operations, as was demonstrated experimentally (Van Dorst and Peters, 1974). Because the first diseased plants were near the sides and then often near fissures and near the main walk of glasshouses, Van Dorst and Peters (1974) suggested that the agent might be introduced by a vector, supposedly an insect. Results of tests to determine whether a common aphid species (*Myzus persicae*) was able to transmit the agent were, however, negative. Neither was there any evidence that the disease was soil borne or seed transmitted (Van Dorst and Peters, 1974).

2.6.5 Natural Hosts

The disease has been reported to occur only in several cultivars of cucumber (*Cucumis sativus* L.).

2.6.6 Strains

So far, no evidence for the existence of more than one strain has been reported.

2.6.7 Economic Losses

Although the pale fruit disease has been detected in many glasshouses in various parts of the Netherlands, the number of infected plants is mostly small, usually less than 20 plants in a crop of 2500 to 25,000 plants. Some higher incidences, however, have been found, perhaps due to spreading by pruning and other cultivation measures (Van Dorst and Peters, 1974). Nevertheless, so far losses seem to have been relatively small.

2.6.8 Control Measures

Apparently, no special measures have been taken so far to control the disease (Van Dorst and Peters, 1974), but, in light of existing knowledge, these would be the same as those used with the diseases described above.

2.7 COCONUT CADANG-CADANG DISEASE

2.7.1 Historical

According to Kent (1953), the name *cadang-cadang* is a local term in the Bicol region of the Philippines signifying yellowing, growth failure, or running-out of any plant. Because of the importance of the coconut disease in question, however, the term has become restricted to apply to this disease alone.

Although the disease was reported to have been present in one area of the Philippine Islands in 1914 and in another area in 1919 (DeLeon and Bigornia, 1953, quoted by Price, 1971), in neither case was the diagnosis made by a plant pathologist and, according to Price (1971), it was likely that the growers who reported it were actually observing decline due to other causes, such as "wet feet."

Cadang-cadang is known to have occurred about 1926 on San Miguel Island (Ocfemia, 1937; Celino, 1947a), but it was not seen with certainty on the mainland of Luzon until 1937 (Price, 1971). By 1950 it was present in all four provinces of the Bicol Peninsula and had spread northeastward to the island of Catanduanes and southward to the island of Masbate and the northern coast of Samar (Price, 1971). Later, the disease was found to have spread to the southeastern tip of Samar (Price and Bigornia, 1969) and northward into Quezon province (Price, 1958, quoted by Price, 1971). By 1971, the disease had further spread into neighboring provinces (Price, 1971).

With cadang-cadang disease, therefore, more than with any other showed that the disease had destroyed an estimated 5,527,000 coconut trees by 1953 (Calica and Bigornia, 1953, quoted by Kent, 1953). By 1959 the number of diseased trees was estimated to be about 7,900,000 (Bigornia *et al.*, 1960, quoted by Price, 1971). A conservative estimate is that more than 12,000,000 trees were destroyed by cadang-cadang in the Bicol Peninsula alone from 1926 to 1971. Evidently, cadang-cadang is a disease of enormous economic importance in the Philippines and, indeed, it is considered to be the main threat to coconut production there (Randles, 1975).

2.7.8 Control Measures

In the absence of any clear indication of the cause of the disease, Price (1971) took issue with earlier recommendations (Celino, 1947a; Ocfemia, 1950; Kent, 1953) that diseased trees should be removed and burned. Price (1971) quoted unpublished work in which cutting and burning of trees, cutting without burning, and clean cultivation were shown to have no effect on the spread of the disease. Also, according to Price (1971), replanting of groves that had been devastated by cadang-cadang was the most sensible recommendation. Sill *et al.* (1963) made similar recommendations, but suggested that replanting should be with either dwarf-type Tambolilid coconuts (as has been done successfully in the Bicol provinces) or else with nuts selected from trees that had survived the disease.

2.8   HOP STUNT DISEASE

Hop stunt disease was first described by Yamamoto *et al.* (1970, quoted by Sasaki and Shikata, 1977a), who recognized it as a transmissible disease on the basis of graft and sap transmission experiments. The disease has been reported to occur throughout the hop gardens in the northern part of the mainland of Japan (Sasaki and Shikata, 1977a). The disease is characterized by shortened internodes of the main and lateral bines and curling of upper leaves (Fig. 9). Hop (*Humulus lupulus* L.) and *H. japonicus* Sieb. et Zucc. are the only known natural hosts (Sasaki and Shikata, 1977a).

Figure 9.   Hop stunt disease. Left: HSV-free hop plant (*Humulus lupulus* cv. Kirin II). Right: Symptoms on HSV-infected hop plant, showing shortened internodes and curled upper leaves. Courtesy: M. Sasaki, Kirin Brewery Co., Takasaki, Japan; and E. Shikata, Hokkaido University, Sapporo, Japan.

disease of concern here, the location of its original occurrence has been well documented, and there is no reason to believe that the disease had occurred much earlier but was not recognized as such by growers or plant pathologists.

2.7.2 Geographic Distribution

Cadang-cadang disease has been reported to occur only in the Philippine Islands, but a very similar disease, variously called *Guam disease, yellow mottle decline of Guam, cadang-cadang of Guam,* or *Tinangaja,* occurred on Guam (Price, 1971). It is not known whether the two diseases have the same cause.

2.7.3 Symptomatology

The earliest symptoms of cadang-cadang are small, irregularly shaped, bright yellow or orange spots in the leaf lamina. These enlarge and tend to fuse with one another to form streaks, thus giving the old leaves a yellowed or mottled appearance (Price, 1971). The spots differ from similar ones of other causation by the lack of a brown center and by appearing olivacious or water-soaked when viewed by light reflected from the under surface of the leaf or petiole (Price, 1971).

Leaves on diseased trees are not normally distorted, but become brittle and more upright than normal (Fig. 7). As the disease advances, the lower leaves drop more frequently than new leaves are formed, and the crown is reduced to a few short, brittle leaves that are greener than normal though obviously diseased (Price, 1971).

About a year after symptoms are first visible in the leaves, the nuts of some fruit trusses are smaller and more numerous than normal and are distinctly rounded on the stigmatic end rather than triangular (Price, 1971). As the disease advances further, the fruits become still smaller, are elongated or otherwise distorted, and further reduced in number (Fig. 8). Eventually, the tree ceases to fruit (Price, 1971). Female flowers fail to set fruit or do not carry it to maturity; later, trees bear only male flowers on dwarfed inflorescenses. Finally, flowers cease to be formed. Trees in a late stage of the disease may linger for 2 or 3 years with a crown reduced to a few short, brittle, upright leaves, resembling a brush. Eventually, the bud dies and falls off, leaving a bare trunk. Trees usually die 8 to 15 years after the first appearance of symptoms (Price, 1971).

Figure 7. Coconut cadang-cadang disease. Left: Early symptoms of disease. Nuts become smaller, rounded, and are more numerous, normal flowers are still borne by inflorescences, fronds appear normal from the ground. Right: Late stage of disease. Nut, inflorescence, and spathe production has ceased, fewer fronds remain, leaflets (pinnae) have become brittle and have ragged appearance. Fronds appear yellow-green from the ground. The late stage lasts from 1 to 4 years, palms finally die. Courtesy: J. W. Randles, University of Adelaide, Australia.

2.7.4 Field Transmission and Spread

Ocfemia (1937), Celino (1947b), and Reinking (1950) presented evidence from which spread of the disease could be inferred. Thus on San Miguel Island only a few trees at the northwestern tip of the island showed symptoms of cadang-cadang originally. The disease then spread slowly in a southeasterly direction until in 1962 only 80 healthy palms of an original 250,000 remained (Price, 1971).

Bigornia *et al.* (1960, quoted by Price, 1971) studied spread of the disease in several large plots. They reported that the spread in these plots was slow and random. The randomness of spread was confirmed by Price and Bigornia (1969), who concluded that the spread was as

Figure 8. Coconut cadang-cadang disease. Top: Nuts from healthy (left) and diseased palms; note reduction in size, rounding and longitudinal surface scarifications associated with disease. Bottom: Yellow spots on the leaflets of a seedling inoculated with nucleic acids from diseased palms. The yellow spots are characteristic of the disease as seen in older, naturally infected palms. Courtesy: J. W. Randles, University of Adelaide, Australia.

though the disease was transmitted by a vector from a source outside the plot.

Holmes (1961a, b) came to similar conclusions and further observed that, according to his plotted data on spread, the putative disease agent did not spread from coconut to coconut, but only from a weed reservoir. He implicated a weed commonly occurring in coconut plantations, *Elephantopus mollis*, as the reservoir. Nagaraj (1967, quoted by Price, 1971) supported Holmes' hypothesis of a weed reservoir, but he did not agree that the principal weed was *E. mollis*. Contrary to Holmes' conclusions, Price and Bigornia (1972) presented evidence for the spread of the disease from coconut to coconut tree.

Evidently, despite the clear evidence of field spread of cadang-cadang disease, the mechanism(s) by which this occurs is still unknown.

2.7.5   Natural Hosts

In addition to the coconut palm (*Cocos nucifera* L.), cadang-cadang apparently attacks other species, as judged by the symptoms they display. Among these, the African oil palm (*Eleais guineensis* Jacq.) is even more severely affected than the coconut palm (Price, 1971). Symptoms similar to those of cadang-cadang have been observed also in the Buri or corypha palm (*Corypha elata* Roxb.), in the anahaw palm [*Livistonia rotundifolia* (Lam.) Mart.], in the buñga or betel palm (*Areca catechu* L.), in the buñga de China (*Adonidia merrillii* Becc.), in the pungahan or fish tail palm (*Caryota cummingii* Lodd.), and possibly in *Pandanus coplandii* Merrill (Kent, 1953; Price, 1971).

Although, as mentioned above, the existence of natural reservoirs of the putative agent in certain weed species has been suspected, no conclusive evidence exists whether this is, in fact, the case.

2.7.6   Strains

No strains deviating from the one described appear to have been reported.

2.7.7   Economic Losses

In 1957, losses in yield of copra due to cadang-cadang disease had been estimated to be about $16,646,000 annually (Price, 1971). Strip surveys showed that the disease had destroyed an estimated 5,527,000 coconut trees by 1953 (Calica and Bigornia, 1953, quoted by Kent, 1953). By 1959 the number of diseased trees was estimated to be about 7,900,000 (Bigornia et al., 1960, quoted by Price, 1971). A conservative estimate is that more than 12,000,000 trees were destroyed by cadang-cadang in the Bicol Peninsula alone from 1926 to 1971. Evidently, cadang-cadang is a disease of enormous economic importance in the Philippines and, indeed, it is considered to be the main threat to coconut production there (Randles, 1975).

2.7.8 Control Measures

In the absence of any clear indication of the cause of the disease, Price (1971) took issue with earlier recommendations (Celino, 1947a; Ocfemia, 1950; Kent, 1953) that diseased trees should be removed and burned. Price (1971) quoted unpublished work in which cutting and burning of trees, cutting without burning, and clean cultivation were shown to have no effect on the spread of the disease. Also, according to Price (1971), replanting of groves that had been devastated by cadang-cadang was the most sensible recommendation. Sill et al. (1963) made similar recommendations, but suggested that replanting should be with either dwarf-type Tambolilid coconuts (as has been done successfully in the Bicol provinces) or else with nuts selected from trees that had survived the disease.

2.8 HOP STUNT DISEASE

Hop stunt disease was first described by Yamamoto et al. (1970, quoted by Sasaki and Shikata, 1977a), who recognized it as a transmissible disease on the basis of graft and sap transmission experiments. The disease has been reported to occur throughout the hop gardens in the northern part of the mainland of Japan (Sasaki and Shikata, 1977a). The disease is characterized by shortened internodes of the main and lateral bines and curling of upper leaves (Fig. 9). Hop (Humulus lupulus L.) and H. japonicus Sieb. et Zucc. are the only known natural hosts (Sasaki and Shikata, 1977a).

Figure 9.   Hop stunt disease. Left: HSV-free hop plant (*Humulus lupulus* cv. Kirin II). Right: Symptoms on HSV-infected hop plant, showing shortened internodes and curled upper leaves. Courtesy: M. Sasaki, Kirin Brewery Co., Takasaki, Japan; and E. Shikata, Hokkaido University, Sapporo, Japan.

3. EXPERIMENTAL BIOLOGY

In this chapter, we shift our attention from observation of the naturally occurring viroid diseases in the field to experimental work on the biological characteristics of these diseases. With each disease, these biological studies laid the groundwork for the later identification of the causative agent as a viroid. Because most of the experiments considered in this chapter were done in the greenhouse, we might say that we are moving from the field (Chapter 2) through the greenhouse (this chapter) into the laboratory (later chapters).

3.1 PROPAGATION OF VIROIDS

To study the properties of the causative agent of an infectious disease, one obviously must have a ready source of the agent. Although this source could consist of infected plants of the same species in which the agent is found in the field, such as potatoes in the case of potato spindle tuber disease or certain citrus trees in the case of citrus exocortis disease, this approach sometimes is not practical because of long times of incubation, relatively low titer of the viroid, or undesirable plant constituents.

For these and other reasons, efforts have been made to find more suitable propagation hosts of viroids and as a corollary of these studies, new experimental hosts were detected in several cases. Also, some attempts have been made to propagate viroids in tissue and cell cultures, as well as in protoplasts.

3.1.1   Experimental Host Plants and Host Ranges

For each known viroid, the experimentally determined host range is
listed, as well as the preferred hosts for propagation of the viroid.

3.1.1.1   Potato Spindle Tuber Viroid

Systematic work on the properties of the causative agent of this disease
became feasible only after Raymer and O'Brien (1962) had discovered
that the agent is able to infect tomato (*Lycopersicon esculentum* Mill.)
and to produce in certain cultivars of this host a characteristic syndrome
that cannot be confused readily with that of any other known potato
or tomato pathogen (except bunchy top of tomato; see Section 3.1.1.2). In
the cultivar Rutgers, first symptoms consist of epinasty, rugosity, and lateral twisting of leaflets formed after inoculation (Fig. 10). Later, necrosis
of midribs and lateral veins of these rugose leaflets becomes evident
(Raymer and O'Brien, 1962). This necrosis of the vascular tissue
gradually becomes more extensive and causes the leaves at three to five

Figure 10.   Healthy (right) and PSTV-infected *Lycopersicon esculentum* (tomato) cv.
Rutgers plants. Note stunting and epinasty in infected plant.

nodes to become first yellow and then brown. Small, mildly rugose leaves with less severe vascular necrosis and epinasty continue to develop above the necrotic region. At this stage, lower leaves of the plants are normal in appearance. Infected plants are severely stunted when compared with healthy ones. No apparent interference with flower or fruit development occurs, except that the fruits are considerably smaller than those produced on healthy plants (Raymer and O'Brien, 1962).

When PSTV was transmitted from infected tomato plants to Saco potatoes, these developed tubers with typical symptoms of spindle tuber, indicating that the spindle tuber viroid had been transmitted to tomato. Inoculation of appropriate indicator plants, as well as serological tests, showed that potato viruses X, A, Y, S, M, and yellow dwarf were absent in the infected tomato plants.

Later tests showed that the viroid concentration in Rutgers tomato plants is comparable to that in infected potato plants (Raymer *et al.*, 1964). Thus this plant is well suited as a propagation host for PSTV (as well as for bioassay; see Section 3.4.1).

Later, many other hosts of PSTV were discovered (Easton and Merriam, 1963; O'Brien and Raymer, 1964; Singh, 1970a; Singh and O'Brien, 1970; O'Brien, 1972; Singh, 1973). In many of these, no symptoms developed and the fact that PSTV had infected these species and replicated could be determined only by back inoculation to tomato or potato, in which the typical symptoms of the disease again developed.

In Table 1, species in which PSTV has been shown to replicate are listed. Singh (1973) assembled a list of 94 species in 31 families that were found to be resistant to PSTV. Interestingly, these include many solanaceous species, some of which belong to genera that include susceptible species, such as *Capsicum*, *Datura*, and *Solanum*.

3.1.1.2 Tomato Bunchy Top Viroid (TBTV)

Raymer and O'Brien (1962) noted the similarity of symptoms induced in tomato by PSTV and those typical of the bunchy top disease of tomato (McClean, 1931). Later, Raymer and O'Brien (see Benson *et al.*, 1965) reported that in comparative studies with the bunchy top "virus" from South Africa and PSTV from the U.S.A., the two "viruses" produce the same symptoms in both tomato and Saco potato and that both are very similar in their thermal inactivation "points" and longevity *in vitro*. On the basis of these similarities, the two agents were considered

Table 1. Suscepts of Potato Spindle Tuber Viroid

Family and Species	Symptoms[a]	References[b]
Amaranthaceae		
Gomphrena globosa L.	−	3
Boraginaceae		
Myosotis sylvatica Hoffm.	−	8
Campanulaceae		
Campanula medium L.	−	8
Caryophyllaceae		
Cerastium tomentosum L.	−	8
Dianthus barbatus L.	−	8
Compositae		
Gynura aurantiaca (B.1) DC	+	8
Convolvulaceae		
Convolvulus tricolor L.	−	8
Dipsaceae		
Scabiosa japonica Miq.	−	8
Nolanaceae		
Nolana sp. "Lavender Gown"	−	7
Sapindaceae		
Cardiospermum halicacabum L.	−	8
Scrophulariaceae		
Antirrhinum sp.	−	7
Diascia barberae Hook. f.	−	7
Nemesia foetens Vent.	−	7
Nemesia sp. "Carnival"	−	7
Penstemon richardsonii Dougl.	−	8
Penstemon sp.	−	7
Solanaceae		
Atropa bella-donna L.	−	8
Browallia demissa L. = *B. Americana* L.	−	8
Browallia grandiflora Grah.	−	8
Browallia speciosa Hook.	−	7
Browallia viscosa H.B.K.	−	8
Capsicum annuum L.	−	3
Capsicum frutescens L. = *C. microcarpum* DC.	−	8
Capsicum nigrum Willd.	−	8
Cyphomandra betacea (Cav.) Sendtn.	−	8
Datura metel L.	−	3
Datura stramonium L.	−	3
Lycopersicon esculentum Mill.	+	1
Lycopersicon glandulosum C.H. Mull.	+	5
Lycopersicon hirsutum Hum. + Bonpl. in Dun.	+	5
Lycopersicon peruvianum (L.) Mill.	+	5

(continued)

Table 1 (continued)

Family and Species	Symptoms[a]	References[b]
Solanaceae (continued)		
Lycopersicon pimpinellifolium (L.) Mill.	−	8
Nicandra physalodes (L.) Pers.	−	3
Nicotiana alata Link & Otto	−	8
Nicotiana bigelovii (Torr.) Wats.	−	8
Nicotiana bonariensis Lehm.	−	8
Nicotiana chinensis Fisch. ex Lehm.	−	8
Nicotiana clevelandii Gray	−	8
Nicotiana debneyi Domin.	−	3
Nicotiana glauca Grah.	−	8
Nicotiana glutinosa L.	+	3
Nicotiana goodspeedii Wheeler	−	8
Nicotiana clevelandii X		
Nicotiana glutinosa Christie	−	8
Nicotiana knightiana Goodspeed	−	8
Nicotiana langsdorffii Weinm.	−	8
Nicotiana longiflora Cav.	−	8
Nicotiana megalosiphon Heurck. & Meull.	−	8
Nicotiana nudicaulis Wats.	−	8
Nicotiana paniculata L.	−	8
Nicotiana plumbaginifolia Viv.	−	8
Nicotiana quadrivalvis Pursh	−	8
Nicotiana raimondii Macbride	−	8
Nicotiana repanda Willd. ex Lehm.	−	8
Nicotiana rotundifolia Lindl.	−	8
Nicotiana rustica L.	−	3
Nicotiana X *sanderae* Hort. ex W. Wats	−	8
Nicotiana solanifolia Walp.	−	8
Nicotiana sylvestris Speg. & Comes	−	8
Nicotiana tabacum L.	−	3
Nicotiana texana Hort. Par ex Maxim.	−	8
Nicotiana viscosa Lehm.	−	8
Nierembergia coerulea Sealy	−	8
Petunia axillaris (Lam.) B.S.P. =		
P. nyctaginiflora Juss.	−	8
Petunia hybrida Vilm.	−	3
" " "	+	8
Petunia inflata Fries	−	8
Petunia violacea Lindl.	−	8
Physalis alkekengi L.	−	8
Physalis angulata L.	−	8
Physalis floridana Rydberg = *P. pubescens* L.	−	3

(continued)

Table 1 (continued)

Family and Species	Symptoms[a]	References[b]
Solanaceae (continued)		
Physalis franchetii Mast. = *P. alkekengi*		
L. var. *franchetii* (Mast.) Mak	−	8
Physalis heterophylla Nees.	−	8
Physalis ixocarpa Brot. ex Hornem.	−	8
Physalis minima L.	−	8
Physalis parviflora Hort. ex Zuccagni	−	8
Physalis peruviana L. = *P. edulis* Sims	−	3
Physalis philadelphica Lam.	−	8
Physalis pruinosa L.	−	8
Physalis somnifera L.	−	8
Physalis viscosa L.	−	8
Salpiglossis sinuata Ruiz & Pav. = *S. variabilis* Hort.	−	8
Salpiglossis spinescens Clos.	−	8
Saracha jaltomata Schlecht.	−	8
Saracha umbellata DC	−	8
Schizanthus pinnatus Ruiz & Pav.	−	8
Schizanthus retusus Hook.	−	8
Scopolia anomala (Link et. Otto) Airy-Schaw.	+	8
Scopolia corniolica Jacq.	+	8
Scopolia lurida Dun.	+	6, 8
Scopolia physaloides Dun.	−	8
Scopolia sinensis Hemsl.	+	6, 8
Scopolia stramonifolia (Wall.) Semenova	+	8
Scopolia tangutica Maxim.	+	8
Solanum aethiopicum L.	−	8
Solanum alatum Moench	−	8
Solanum americanum Mill.	−	8
Solanum atriplicifolium Gill. ex Nees	−	8
Solanum auriculatum Ait. = *S. mauritianum* Scop.	−	8
Solanum aviculare G. Forst.	+	6, 8
Solanum berthaultii Hawkes	+	2, 5
Solanum boliviense Dun.	+	2, 5
Solanum bonariense L.	−	8
Solanum bulbocastanum Dun.	+	2
Solanum capsicastrum Link. ex Schau.	−	8
Solanum carolinense L.	−	8
Solanum cervantesii Lag.	−	8
Solanum chlorocarpum Schur	−	8
Solanum ciliatum Lam.	−	8
Solanum cornutum Lam.	−	8
Solanum decipiens Opiz	−	8

(continued)

Table 1 (continued)

Family and Species	Symptoms[a]	References[b]
Solanaceae (continued)		
Solanum depilatum Kitag.	+	6, 8
Solanum diflorum Vellozo	−	8
Solanum dulcamara L.	−	8
Solanum famatinae Bitt. & Wittm. =		
S. *spegazzinii* Bitt.	+	2, 5
Solanum goniocalyx Juz. & Buk. = *S. stenotonum*		
Juz. & Buk. subsp. *goniocalyx* (Juz. & Buk.) Hawkes	+	2, 5
Solanum gracile Otto ex W. Baxt.	−	8
Solanum guineense L.	−	8
Solanum hendersonii Hort. ex W. F. Wight	−	8
Solanum hibiscifolium Rusby	−	8
Solanum humile Bern. ex Willd.	−	8
Solanum judaicum Bess.	−	8
Solanum kitaibelii Schult.	−	8
Solanum kurtzianum Bitt. & Wittm.	+	2, 5
Solanum laciniatum Ait.	−	8
Solanum luteum Mill.	−	8
Solanum maglia Schlechtd.	+	2, 5
Solanum maritimum Meyen ex Nees	−	8
Solanum melongena L.	+	7
Solanum memphiticum J. F. Gmel.	−	8
Solanum miniatum Bernh. ex Willd.	−	8
Solanum nigrum L.	−	8
Solanum nitidibaccatum Bitt.	−	8
Solanum nodiflorum Jacq.	−	8
Solanum ochroleucum Bast.	−	8
Solanum olgae Pojark.	−	8
Solanum ottonis Hyl.	−	8
Solanum papita Rydb.	−	8
Solanum paranense Dusen	−	8
Solanum persicum Willd. ex Roem. & Schult.	−	8
Solanum polyadenium Greenm.	+	2, 5
Solanum polytrichon Rydb.	+	2, 5
Solanum pseudo-capsicum L.	−	8
Solanum pyracanthum Jacq.	−	8
Solanum rantonnettii Carr. ex Lescuy.	−	8
Solanum rostratum Dunal.	+	4
Solanum saponaceum Dun.	−	8
Solanum sinaicum Boiss.	−	8
Solanum sisymbrifolium Lam.	−	8
Solanum sodomeum L.	−	8

(continued)

Table 1 (continued)

Family and Species	Symptoms[a]	References[b]
Solanaceae (continued)		
Solanum surattense Burm. f.	−	8
Solanum tomentosum L.	−	8
Solanum tuberosum L.	+	−
Solanum tripartitum Dun.	−	8
Solanum umbellatum Mill.	−	8
Solanum verbascifolium L.	−	8
Solanum verneii Bitt. & Wittm.	+	2, 5
Valerianaceae		
Valeriana officinalis L.	−	8

[a] + = Macroscopically visible symptoms.

− = No symptoms, but viroid recoverable by transfer to tomato.

[b] *References.* 1 = Raymer and O'Brien (1962); 2 = Easton and Merriam (1963); 3 = O'Brien and Raymer (1964); 4 = Singh and Bagnall (1968); 5 = Singh and O'Brien (1970); 6 = Singh (1970a); 7 = O'Brien (1972); 8 = Singh (1973).

identical or closely related strains of the same "virus" (Benson *et al.*, 1965). Recent results with PSTV and CEV (which also resemble one another closely and were thought by some to be identical—see Chapter 8) revealed, however, that despite close similarity of the biological properties of two viroids (such as identical host range and type of symptoms produced in different hosts), they may be distinct entities chemically (see Chapter 8). For this reason, TBTV is treated here separately from PSTV.

McClean (1935, 1948) investigated the host range of TBTV. He found that, in a number of species, mostly of the family Solanaceae, the agent is carried without the appearance of symptoms of the infection. These species are listed in Table 2.

Evidently, there is much overlap among species susceptible to PSTV and TBTV. Some apparent discrepancies, however, exist. Thus *Zinnia elegans* has been reported as a susceptible species for TBTV (Table 2), but as a resistant species for PSTV (O'Brien and Raymer, 1964).

3.1.1.3 *Citrus Exocortis Viroid*

With the exocortis disease, also, it was essential to find hosts different from those naturally affected, because the latter are not readily adaptable for greenhouse work with large numbers of plants and because symptoms usually appear only after an extended incubation period.

Table 2. Suscepts of Tomato Bunchy Top Viroid

Family and Species	Symptoms[a]	References[b]
Compositae		
Zinnia elegans Jacq.	−	2, 3
Solanaceae		
Capsicum annuum L.	−	2
Datura stramonium L.	−	3
Lycopersicon esculentum Mill.	+	1
Lycopersicon pimpinellifolium (L.) Mill	−	2
Nicandra physaloides Gaertn.	±	2, 3
Nicotiana glutinosa L.	±	3
Nicotiana tabacum L.	−	2, 3
Petunia hybrida Hort.	−	2, 3
Physalis angulata L.	−	3
Physalis peruviana L.	±	2, 3
Physalis viscosa L.	±	2
Solanum aculeastrum Dunal.	−	2
Solanum aculeatissimum Jacq.	±	2
Solanum ciliatum Lam.	−	2
Solanum duplosinuatum Klotsch.	±	2, 3
Solanum giganteum Jacq.	±	3
Solanum incanum L.	±	2, 3
Solanum indicum L.	−	3
Solanum melongena L.	−	2, 3
Solanum nigrum L.	−	2
Solanum panduraeforme E. M.	±	2
Solanum sisymbrifolium Lam.	−	3
Solanum sodomaeum L.	−	2, 3
Solanum tuberosum L.	−	2

[a] − = No visible symptoms.
± = Slight symptoms, often only under certain growing conditions.
+ = Readily visible symptoms.
[b] *References.* 1 = McClean (1931); 2 = McClean (1935); 3 = McClean (1948).

Weathers (1965a, b) reported transmission of CEV to certain *Petunia* and *Physalis* species. Later, Weathers and Greer (1968, 1972) reported transmission of CEV to two composite species, *Gynura aurantiaca* and *G. sarmentosa*. In *G. aurantiaca*, inoculation with CEV results in symptoms that are similar to those in citron but develop more quickly, usually within 14 days of inoculation. They consist of epinasty of young developing leaves and discoloration and necrosis on the undersides of midveins and main lateral veins (Fig. 11). Necrosis often extends into the petioles

Figure 11. Healthy (right) and CEV-infected *Gynura aurantiaca* plants. Note epinasty of young leaves and discoloration and necrosis on the undersides of midveins and main lateral veins. Courtesy: L. K. Grill and J. S. Semancik, University of California, Riverside, Ca.

and stems, but plants do not die. In later stages, the leaves curl downward into a "roll" and some leaves become malformed and develop irregular green blotches intermixed with the normal purple color. The plants become stunted and develop a "bushiness" because of the production of an abnormal number of secondary shoots. Also, clearing of principal veins and malformation of flowers develop (Weathers and Greer, 1972).

Gynura aurantiaca has become the favored propagation host for

CEV because symptoms develop after a relatively short incubation period, the viroid reaches a high titer which is maintained for long periods, and because the plants are durable, succulent, compact, and are easy to grow and propagate (Weathers and Greer, 1972).

In Table 3, known hosts of CEV are listed.

3.1.1.4 Chrysanthemum Stunt Viroid

The experimental host range of CSV is listed in Table 4. No susceptible species outside of the composite family have so far been identified. As with other viroids, CSV is able to replicate in many species without causing obvious symptoms of disease.

Favored propagation and index host for CSV is the chrysanthemum cultivar Mistletoe [actually a group of cultivars (Brierley, 1953)] because it is one of few chrysanthemum cultivars that develop, in addition to

Table 3. Suscepts of Citrus Exocortis Viroid

Family and Species	Symptoms[a]	References[b]
Compositae		
Gynura aurantiaca (Bl.) DC	+	8
Gynura sarmentosa (Bl.) DC =		
G. *procumbens* (Lour.) Merr.	+	8
Rutaceae		
Citrus aurantifolia (Christm.) Swingle	−	6
Citrus aurantium L.	−	9
Citrus excelsa Wester	−	6
Citrus jambhiri Lush.	−	6
Citrus limetta Risso	+	5
Citrus limettioides Tanaka	+	5
Citrus limon Burm. f.	−	6
Citrus limonia Osbeck	+	3
Citrus medica L.	+	6
Citrus paradisi Macf.	−	3
Citrus reticulata Blanco	−	4
Citrus sinensis (L.) Osbeck	−	4
Citrus sinensis X *Poncirus trifoliata*	+	2
Poncirus trifoliata (L.) Raf.	+	1
Solanaceae		
Lycopersicon esculentum Mill.	+	10
Petunia axillaris (Lam.) B.S.P.	+	7
Petunia hybrida Vilm.	+	7

(continued)

Table 3.   (continued)

Family and Species	Symptoms[c]	References[b]
Petunia violacea Lindl.	+	7
Physalis floridana Rydberg	+	7
Physalis ixocarpa Brot. ex Hornem.	+	7
Physalis peruviana L.	+	7
Scopolia carniolica Jacq.	+	11
Scopolia lurida Dun.	+	11
Scopolia physaloides Dun.	−	11
Scopolia sinensis Hemsl.	+	11
Scopolia stramonifolia (Wall.) Semenova	+	11
Scopolia tangutica Maxim.	+	7
Solanum aculeatissimum Jacq.	+	7
Solanum dulcamara L.	+	7
Solanum hispidum Pers.	+	7
Solanum integrifolium Poir.	+	7
Solanum marginatum L. f.	+	7
Solanum quitoense Lam.	+	7
Solanum topiro Humb. & Bonpl. ex Dunal	+	7
Solanum tuberosum L.	+	11

[a] − = No visible symptoms, but viroid recoverable.
   + = Visible symptoms.
[b] *References.* 1 = Benton *et al.* (1949); 2 = Bitters *et al.* (1954); 3 = Olson and Shull (1956); 4 = Calavan *et al.* (1959); 5 = Weathers and Calavan (1961); 6 = Calavan *et al.* (1964); 7 = Weathers *et al.* (1967); 8 = Weathers and Greer (1968, 1972); 9 = Garnsey and Whidden (1970); 10 = Semancik and Weathers (1972c); 11 = Singh *et al.* (1972).

severe stunting of plants, diagnostic chlorotic spots on the leaves of infected plants (Keller, 1951) (Fig. 4a).

3.1.1.5   *Chrysanthemum Chlorotic Mottle Viroid*

Various cultivars of *Chrysanthemum morifolium* and *Chrysanthemum Zawadskii* var. *latilobum* are the only known hosts of ChCMV (Horst and Romaine, 1975). Commonly used hosts for other viroids were tested, but they were found to be insusceptible for ChCMV (Horst and Romaine, 1975). Individual chrysanthemum cultivars greatly vary in their response to infection. Some cultivars, such as Yellow Delaware, Delaware, Blue Ridge, Deep Ridge, and Tinker Bell, develop marked mottling, followed by general chlorosis; others, such as Dark Red Star, Giant Betsy Ross, Hurricane, and Icecap, also develop marked mottling and chlorosis, but revert later to normal or near-normal vigor and

Table 4. Suscepts of Chrysanthemum Stunt Viroid

Family and Species	Symptoms[a]	References[b]
Compositae		
Achillea millefolium L.	−	3
Achillea ptarmica L.	−	2, 3
Ambrosia trifida L.	−	3
Anthemis tinctoria L.	−	2, 3
Centaurea cyanus L.	−	2, 3
Chrysanthemum carinatum Schousb.	−	2, 3
Chrysanthemum cinerariaefolium (Trevar.) Vis.	−	3
Chrysanthemum coccineum Willd.	+	2, 3
Chrysanthemum coronarium L.	−	2, 3
Chrysanthemum corymbosum L.	−	3
Chrysanthemum frutescens L.	−	3
Chrysanthemum hortorum = *C. morifolium*	+	2
Chrysanthemum lacustre Brot.	−	3
Chrysanthemum leucanthemum L.	−	3
Chrysanthemum majus (Desf.) Aschers.	−	3
Chrysanthemum maximum Ramond	−	3
Chrysanthemum morifolium (Ramat.) Hemsl.	+	1
Chrysanthemum myconis L.	−	3
Chrysanthemum nivellei Braun-Blanq. & Maire	−	3
Chrysanthemum parthenium (L.) Pers.	+	3
Chrysanthemum parthenium f. *flosculosum* (DC) Beck.	+	3
Chrysanthemum praealtum Vent.	+	2, 3
Chrysanthemum sp.	−	3
Chrysanthemum viscosum Desf.	−	3
Dahlia pinnata Cav.	−	3
Dahlia variabilis Desf.	−	2, 3
Echinacea purpurea (L.) Moench	−	2, 3
Emilia sagittata (Vahl) DC	−	2, 3
Heliopsis pitcheriana Hort.	−	3
Liatris pycnostachya Michx.	−	3
Liatris spicata Willd.	−	3
Sanvitalia procumbens Lam.	−	3
Senecio cruentus DC.	+	2, 3
Senecio glastifolius L. f.	+	3
Senecio mikanioides Otto	−	3
Sonchus asper (L.) Hill	−	5
Tanacetum boreale Fisch.	−	3
Tanacetum camphoratum Less.	−	3
Tanacetum vulgare L.	−	3
Tithonia rotundifolia (Mill.) Blake	−	3, 4
Venidium fastuosum (Jacq.) Stapf.	−	3
Verbesina encelioides (Cav.) B. & H.	−	3
Zinnia elegans Jacq.	−	3

[a] *References*. 1 = Dimock (1947); 2 = Brierley (1950); 3 = Brierley (1953); 4 = Keller (1953); 5 = Hollings and Stone (1973).

color; still others, such as Knob Hill, Matador, Mermaid, and Red
Cap, only develop chlorotic spotting, vein clearing, and very mild
chlorosis; and a large number of cultivars, such as Albatross, Blue
Chip, Bright Golden Anne, and Fanfare, remain symptomless (Dimock
et al., 1971).

As a propagation and index host, the cultivar Deep Ridge has been
used most often (Romaine and Horst, 1975).

3.1.1.6   Cucumber Pale Fruit Viroid

The known host range of CPFV is restricted almost exclusively to the
family Cucurbitaceae (Van Dorst and Peters, 1974) (Table 5).

The only noncucurbitaceous host reported so far is tomato (Sänger
et al., 1976), and the tomato cultivar Rentita has been used as propagation host (Sänger *et al.*, 1976).

3.1.1.7   Coconut Cadang-Cadang

This disease agent has been propagated only in coconut palm (Randles
et al., 1977). Its viroid nature has not been unequivocally established.

3.1.1.8   Hop Stunt Viroid

This viroid has been recognized only recently (Sasaki and Shikata, 1977a,
b). So far it has been propagated only in hops (*Humulus lupulus* L.),
but some experiments indicate that the causative viroid can be transmitted into several cucurbitaceous species, and into *Humulus japonicus*
Sieb. et Zucc. (Sasaki and Shikata, 1977a). In cucumber, symptoms were
similar to those incited by CPFV. No results of back-transfers to hop
plants have, however, been reported, and the identity of the symptom-
causing agents is, therefore, not definitely established (Sasaki and
Shikata, 1977a).

3.1.2   Environmental Conditions

3.1.2.1   Temperature

Already Raymer and O'Brien (1962) recognized the importance of
environmental conditions for the propagation and symptom development

Table 5. Suscepts of Cucumber Pale Fruit Viroid

Family and Species	Symptoms	Plant Death	References[a]
Cucurbitaceae			
Benincasa cerifera Savi	+	+	1
Benincasa hispida (Thunb.) Cogn.	+	+	1
Bryonopsis laciniosa Naud.	+	+	1
Cayaponia africana (Hook.f.) Exell	−		1
Citrullus Colocynthis Schrad.	+		1
Citrullus vulgaris Schrad.	+		1
Coccinia sessilifolia (Sond.) Cogn.	−		1
Cucumeropsis edulis (Hook.f.) Cogn.	+	+	1
Cucumis africanus L.f.	−		1
Cucumis Anguria L.	+		1
Cucumis dipsaceus Ehrenb.	+		1
Cucumis Melo L.	+	+	1
Cucumis metuliferus E. Mey.	+	+	1
Cucumis myriocarpus Naud.	+		1
Cucumis sativus L.	+		1
Cucurbita andreana Naud.	−		1
Cucurbita ficifolia Bouché	−		1
Cucurbita maxima Duchesne	−		1
Cucurbita mixta Pangalo	−		1
Cucurbita moschata Duchesne	−		1
Cucurbita Pepo L.	−		1
Kedrostis africana (L.) Cogn.	−		1
Langenaria vulgaris Ser.	+	+	1
Luffa acutangula Roxb.	+		1
Luffa cylindrica (L.) Roem.	+		1
Luffa operculata (L.) Cogn.	+		1
Melothria japonica (Thunb.) Maxim.	+		1
Melothria pendula L.	−		1
Melothria scabra Naud.	−		1
Trichosanthes Anguina L.	+		1
Solanaceae			
Lycopersicon esculentum Mill.	+		2

[a] References. 1 = Van Dorst and Peters (1974); 2 = Sänger et al. (1976).

of PSTV in tomato. They reported that at Beltsville, Maryland, only 10 days were required for symptom expression in early fall, but that this period was extended to as long as 42 days during November through January. The authors commented that this variation may account in part for the failure of earlier workers (Goss, 1930a) to transmit PSTV to tomato. In Raymer and O'Brien's (1962) experiments, plants were grown at greenhouse temperatures of about 26°C during the day and 21°C at night, except during June through September, when the day temperatures ranged from about 32°C to 40°C. Later, it became evident that PSTV reaches far higher titers and symptoms are expressed earlier when plants are kept at relatively high temperatures, such as 30°C to 35°C, than at lower temperatures (Singh and O'Brien, 1970).

CEV similarly reaches far higher titers and symptoms are expressed far earlier when inoculated *Gynura aurantiaca* plants are kept at or above 30°C rather than at 20°C (Sänger and Ramm, 1975).

With CSV, on the other hand, symptom expression is optimal at 21°C (with high light intensity) and masking of symptoms occurs above 30°C even with high light intensity (Hollings and Stone, 1973).

With ChCMV, symptom expression is most pronounced and occurs earlier at day/night temperatures of 21.1°C/15.5°C than at either higher or lower temperatures (Dimock *et al.*, 1971).

High temperature (30°C to 32°C) favors symptom development of CPFV propagated in cucumber plants by drastically shortening the incubation period and increasing severity of symptoms (Van Dorst and Peters, 1974).

HSV propagated in cucumber plants expresses symptoms earlier when plants are kept at about 30°C than when kept at about 25°C (Sasaki and Shikata, 1977a).

3.1.2.2 Light

The effect of light on symptom expression has been studied most carefully wih CSV in chrysanthemum. Brierley *et al.* 1952) noted that Mistletoe chrysanthemums systemically infected with CSV express yellow leaf spotting in summer, whereas during winter such plants develop fewer and smaller spots, or none, even if greenhouse daylight is supplemented by 6 hours at night from 100-W incandescent lamps at an intensity of about 30 to 60 footcandles at the leaf surface. Symptom expression was improved after 3 weeks by substituting 300-W lamps; and after only two 16-hour days under a completely artificial source supplying about 2000 footcandles at the leaf surface, symptoms became clearly defined. After 3 to 12 days in this light a gradual increase in the degree of leaf distortion and in the ratio of yellow-to-green areas occurred (Brierley *et al.*, 1952). Similar observations on the effect of light on symptom expression of CSV were made by Teyssier and Dunez (1971) and by Hollings and Stone (1973).

Although the effect of light intensity on titer and symptom development in PSTV-infected tomato and CEV-infected *Gynura aurantiaca* has not been studied systematically, experience indicates that high light intensity favors both high titer and early symptom expression (Raymer *et al.*, 1964; Sänger and Ramm, 1975).

McClean (1931) reported that high light intensity (and high temperature) lead to an intensification of symptoms of TBTV in tomato.

With ChCMV also, high light intensity appears to be advantageous; Romaine and Horst (1975) used 2000 footcandles of fluorescent and incandescent illumination for the propagation of ChCMV in the chrysanthemum cultivar Deep Ridge.

Van Dorst and Peters (1974) used supplementary illumination during winter months in their work with CPFV.

3.1.2.3 Host Nutrition

With PSTV, application of sufficient fertilizer to insure vigorous growth of plants appears to favor early symptom formation and maximal viroid titer (O'Brien and Raymer, 1964). In the tomato cultivar Allerfrüheste-Freiland, dramatic increases in the characteristic veinal necrosis symptoms induced by the mild or severe strain of PSTV occurred when the level of manganese in the growth medium was increased from 0 to 9 μg/ml (Lee and Singh, 1972).

Weathers *et al.* (1965) studied the effect of host nutrition on symptoms induced by CEV and concluded that nutrition effects in citrus plants depend on the specific host-"virus" complex, and that the "virus" does not always follow the same pattern. No relation between vigor of citrus plants and CEV activity (as measured by development of symptoms) was observed. Conditions that favored the growth of the host did not necessarily favor symptom formation. Conversely, factors that disfavored the host (excess nitrogen and phosphorus) favored symptom expression.

3.1.3 Propagation in Tissue and Cell Cultures and in Protoplasts

In principle, propagation of viroids in plant tissue culture systems is preferable over propagation in intact plants. Plant tissue culture systems, however, still pose problems and it is, therefore, not surprising that few efforts have been made to develop such systems for viroid propagation. Isolated protoplasts from tobacco leaves have been used successfully to study the primary events after inoculation with tobacco mosaic and some other plant viruses (Sakai and Takebe, 1974; Zaitlin and Beachy, 1974) or with plant viral RNA (Aoki and Takebe, 1969). The possibility, therefore, exists that suitable protoplasts derived from host plants of viroids could be infected with viroids and used in an analogous manner.

Mühlbach *et al.* (1977) isolated viable protoplasts from leaves of healthy and viroid-infected tomato plants by a two-step enzyme treatment. Such protoplasts incorporated radioactive uridine into all cellular RNA species. In protoplasts from tomato leaves systemically infected with CEV an extra band of radioactive viroid could be detected 30 hours after adding the [^{3}H]uridine. These results indicate, however, that the rate of viroid synthesis in protoplasts from systemically infected leaves is low as compared with that in intact plants and that the ratio of [^{3}H] incorporation into viroid ranges only between 0.001 and 0.3% of the incorporation into tRNA, whereas it may reach up to about 3.5% in systemically infected intact plants (Mühlbach *et al.*, 1977).

In other experiments, protoplasts isolated from three tomato cultivars were inoculated with three viroids; CEV, PSTV, and CPFV. Both viroid bioassay and [^{3}H]uridine incorporation indicated that CPFV was able to replicate in protoplasts of the cultivar Hilda 72, but not in the cultivars Rentita or Rutgers (Mühlbach and Sänger, 1977). No convincing evidence for CEV or PSTV replication in protoplasts from any of the three cultivars could be obtained. Also, the minimal viroid concentration necessary for infection to occur (10 μg/ml) appears excessive and would limit usefulness of the system.

Evidently, much more work is necessary before plant protoplast systems can become practical alternatives to propagating viroids in intact plants.

3.2 EXPERIMENTAL TRANSMISSION

Much of the early work on plant diseases now recognized as viroid diseases was done in efforts to determine whether the diseases were infectious. It was, therefore, necessary to attempt to transmit a putative agent. Various grafting procedures were used in many early studies. Also, much experimental transmission work was done in efforts to understand the modes of spread of the diseases under study in the field.

Much effort was expended in efforts to identify arthropod species as vectors of the diseases. The fact that viroids usually are transmitted readily by mechanical means was not understood until relatively recently and, in the absence of strict measures to prevent mechanical transmission, results of earlier experiments on other modes of transmission must be interpreted with caution.

3.2.1 Graft Transmission

With the recognition that PSTV is readily transmissible by mechanical means (see Chapter 2), experimental work relied almost exclusively on mechanical transmission of the viroid. Only under special circumstances is grafting still a useful method. Thus O'Brien and Raymer (1964) reported that PSTV is not readily transmitted by sap from infected *Datura stramonium* plants, whereas the viroid is regularly transmitted by side-grafting of infected *D. stramonium* onto tomato plants. In this case, the difficulty encountered in sap inoculation is probably a result of the very high ribonuclease content of *Datura* tissue (Diener and Heinze, 1962).

Citrus exocortis disease basically is a graft union problem and grafting played an important role in the early investigations of the disease, as well as in the natural spread of the disease. As with PSTV, however, all recent work has utilized mechanical transmission.

In contrast to CSV, ChCMV is not readily transmitted by transferring sap from a diseased plant to a healthy one (Horst and Romaine, 1975). Although grafting could be used to overcome this difficulty, improvements in the techniques of mechanical transmission (Romaine and Horst, 1975) have rendered this approach obsolete (see Section 3.2.6).

With other viroids (CPFV and HSV), mechanical transmissibility of the agent was recognized early and grafting has not played a major role in the investigation of the respective diseases. With cadang-cadang disease, finally, grafting is not possible and all work must, by necessity, utilize alternative methods of transmission.

3.2.2 Dodder Transmission

Weathers and Harjung (1964) used dodder (*Cuscuta subinclusa* Dur. and Hilg.) for the transmission of several citrus viruses, but they failed to transmit CEV by this means. Later, however, Weathers (1965b) reported transmission of CEV by dodder from *Citrus aurantifolia* to Etrog citron.

Keller (1953) successfully used another dodder species (*Cuscuta gronovii* Willd.) to transmit CSV from chrysanthemum to chrysanthemum, but Hollings and Stone (1973) were unable to transmit CSV through *Cuscuta campestris* Yuncker. Van Dorst and Peters (1974) were able to transmit CPFV from cucumber to cucumber by use of *Cuscuta subinclusa*, but not via *C. campestris*.

3.2.3 Transmission by Arthropods

As discussed above, a number of workers have suspected that arthropods play a role in the spread of diseases now recognized to be viroid incited, by acting as vectors of the disease agents. No unequivocal demonstration of the transmission of a viroid by an arthropod vector, however, seems to exist.

Early arthropod transmission studies with PSTV have been summarized in Chapter 2. Apparently, no more recent attempts have been made to determine whether arthropods are implicated in the spread of PSTV.

Laird *et al.* (1969) attempted to transmit CEV with the following insects: *Trialeurodes vaporariorum*, the greenhouse whitefly; *Bemisia tabaci*, the sweet-potato whitefly; *Myzus persicae*, the green peach aphid; *Aphis gossypii*, the cotton aphid; *Aphis spiraecola*, the spirea aphid; *Paratrioza cockerelli*, the potato psyllid; and *Circulifer tenellus*, the beet leafhopper. No transmissions were obtained with any of the insects, and the authors reaffirmed the belief that CEV was spread principally through propagation of infected plants or by contaminated grafting tools. The authors considered it unlikely that vectors played any significant role in transmitting CEV, if indeed such vectors existed (Laird *et al.*, 1969).

With the tomato bunchy top agent, McClean (1931) was similarly unable to demonstrate insect transmission.

Brierley and Smith (1949) reported CSV transmission by the aphid *Rhopalosiphum rufomaculatum* (Wilson) to 10 of 35 inoculated chrysanthemum plants and in low and perhaps insignificant proportions by four other aphid species. The significance of these results appears questionable because 6 out of 100 noninoculated control plants also acquired the agent, and later studies indeed showed that no aphid species transmitted CSV (Brierley and Smith, 1951).

Attempts to transmit CPFV with aphids (*Myzus persicae*) were unsuccessful, but the observation that the first diseased plants are found near the sides and then often near fissures and near the main walk of greenhouses suggests that the disease may be introduced by a vector, supposedly an insect (Van Dorst and Peters, 1974). This vector, the authors speculated, may enter the glasshouse along the main walk with carts, material transported, or on workers' clothes. A consequence of these ideas is the assumption that the pathogen occurs in plants and in vectors in the field.

Many experiments were made in efforts to demonstrate insect transmission of coconut cadang-cadang disease, but, despite tests involving more than 100 species of insects, not a single experimental transmission was demonstrated (Price, 1971).

In view of these results, one may venture the generalization that viroid diseases apparently do not rely for their natural spread on transmission by arthropod vectors.

3.2.4   Seed and Pollen Transmission

The earliest evidence indicating vertical transmission of viroids was presented by McClean (1948), who demonstrated that the tomato bunchy top agent is transmitted through the seeds of *Solanum incanum* and *Physalis peruviana*, but not through the seed of tomato, *Nicotiana glutinosa*, or *Physalis viscosa*. Later, Benson and Singh (1964) reported seed transmission of PSTV through the true seed of both potato and tomato, and Hunter *et al.* (1969a) obtained 100% seed transmission of PSTV in the potato cultivars Katahdin and Russet Sebago when both parents were infected. Singh (1970b) presented further evidence for vertical transmission of PSTV. In tomato seedlings derived from crosses where the male parent was infected and the female parent was healthy, PSTV was transmitted in about 9% of the seedlings; where the female parent was infected and the male one was healthy, transmission was about 6%. Higher incidences of transmission (11%) occurred when both parents were infected. A cross involving potato cultivars Saco and Red Pontiac resulted in 12% transmission when both parents were infected, while Saco seed fertilized with Saco pollen yielded about 6% infected seedlings, provided that both parents were infected.

These results were confirmed and extended by Fernow *et al.* (1970), who demonstrated the presence of PSTV in the seed and pollen of diseased potato plants and showed that transmission through the seed from open-pollinated female parents to the seedlings occurred frequently (average 31%), but that it varied in individual collections from zero to 100%. The amount of transmission did not appear to be correlated with variety or with age of the seed (Fernow *et al.*, 1970).

With CEV, on the other hand, evidence concerning seed transmission is more ambiguous. Moreira (1959) and Rossetti (1961) reported negative results, but later Salibe and Moreira (1965b) presented circumstantial evidence that, in their opinion, constituted a strong reason to believe in seed transmission of CEV.

CSV apparently is not seed transmitted (Smith, 1972), and in preliminary experiments with CPFV Van Dorst and Peters (1974) could find no evidence for seed transmission of the viroid.

3.2.5 Soil Transmission

Circumstantial evidence obtained from observation of field spread of viroid diseases does not support the contention that these diseases are soil-borne and limited experimental evidence confirms this conclusion for TBTV (McClean, 1931), PSTV (Goss, 1931), and CPFV (Van Dorst and Peters, 1974).

3.2.6 Mechanical Transmission

All known viroids are transmissible by mechanical means, either readily or with some difficulty. Essentially in all modern work on viroids, mechanical transmission has been used. Some of the earlier methods of mechanical transmission have been described in Chapter 2. Today, three methods only (with several modifications of each) are used almost exclusively. These are described here.

3.2.6.1 Transmission by Leaf Abrasion

This method is identical with standard inoculation procedures used with conventional plant viruses and with nucleic acids isolated from virions. With an appropriate inoculum, this method has been found suitable for transmission of PSTV, TBTV, CSV, ChCMV, CPFV, and HSV. Most workers dust the upper side of leaves to be inoculated with an abrasive, usually carborundum of 500 to 600 mesh, and then lightly rub the leaves with cotton-tipped applicators or cheesecloth pads soaked with inoculum. Usually, leaves are washed with a spray of water immediately after inoculation.

3.2.6.2 Transmission by Knife-Cuts

This method was developed as a consequence of the finding that CEV is readily transmitted with contaminated tools (Garnsey and Jones, 1967). It consists in contaminating a knife or razor blade with CEV by either dipping it into inoculum or by cutting into the stem of an infected plant and then making a number of cuts into the stems or petioles of the receptor plants (Garnsey and Whidden, 1970). With CEV, this procedure proved to be more efficient than needle puncture or leaf abrasion procedures (Garnsey and Whidden, 1973). Often, a combination of knife-cutting and leaf abrasion has been used with CEV (Semancik and Weathers, 1968b) and with CPFV (Van Dorst and Peters, 1974).

3.2.6.3 Transmission by High Pressure Injection

This method has been used with cadang-cadang disease in transmission experiments with the cadang-cadang associated RNA species (Randles et al., 1977). Either a "Hypospray" injector with an electrically operated air compressor or a hand primed "Panjet" mechanical hand injector was used to introduce nucleic acid preparations into coconut palm seedlings.

3.3 IDENTIFICATION OF VIROIDS

Correct identification of viroids is essential in practical work designed to control viroid diseases as well as in work on the physical, chemical, or biological properties of viroids. Because of their unique properties, certain identification procedures that are widely used with conventional plant viruses are not applicable to viroids. This includes identification by electron microscopy, which in the absence of viral nucleoprotein particles is impossible, and, as we shall see, serological identification methods.

3.3.1 Diagnostic Hosts

Some viroids produce such characteristic symptoms in their natural hosts that no different diagnostic host is required for positive identification. This appears to be the case with CPFV in cucumber (Van Dorst and Peters, 1974).

Most viroids, however, incite symptoms in their natural hosts that can be confused with those of other diseases or, more frequently, are indistinct in certain cultivars. In these cases, positive identification must be made by transfer of the pathogen to a suitable diagnostic host. In Table 6, widely used diagnostic hosts are listed for each viroid disease.

With PSTV it was particularly important to find a suitable indicator host because symptoms in some potato cultivars are difficult to recognize and are usually nonexistent in a first-year infection (see Chapter 2), making it difficult to exclude PSTV-infected tubers from next year's "seed" stock.

Although indexing on Rutgers tomato greatly facilitated this task, some workers encountered difficulties in that they were unable to identify all infected plants by indexing on tomato. Thus Fernow (1967) detected only about 13% of the infected clones tested. These poor results were attributed to the existence of a strain that produced symptoms in tomato so mild that they were easily overlooked. To overcome this difficulty, the author developed a double inoculation technique that made possible the ready detection of either the mild or severe strain. With this procedure, PSTV-infected tubers were effectively eliminated from "seed" stocks before planting (Fernow *et al.*, 1969). A further improvement in the indexing procedure was implemented by Singh *et al.* (1970), who used RNA, extracted from PSTV-infected plants by a simplified procedure, as inoculum in Fernow's cross-protection test.

Singh (1970a) reported that the tomato cultivar Allerfrüheste-Freiland reacts with diagnostic symptoms to both the severe and mild strains of PSTV, but Lee and Singh (1972) stated that symptom expression has at times been inconsistent. They determined that veinal necrosis symptoms were critically dependent on adequate concentrations of manganese available to the plants.

Although mild strains of PSTV produce few or no symptoms in Rutgers tomato plants grown in low light intensities during winter months, well-fertilized plants inoculated at an early stage of development and grown under adequate light intensities at reasonably high

Table 6. Diagnostic Viroid Hosts

Viroid	Species	Cultivar	Method of Inoculation	References
PSTV	*Lycopersicon esculentum*	Rutgers	Mechanical	Raymer & O'Brien (1962)
"	"	Allerfrüheste-Freiland	"	Singh (1970a)
"	*Scopolia sinensis*[a]	—	"	Singh (1971)
TBTV	*Lycopersicon esculentum*	Rutgers	"	Benson *et al.* (1965)
CEV	*Citrus medica*	Etrog	Grafting	Calavan *et al.* (1964)
"	*Gynura aurantiaca*	—	Mechanical	Weathers & Greer (1972)
CSV	*Chrysanthemum morifolium*	Mistletoe	"	Keller (1953)
ChCMV	"	Deep Ridge	"	Dimock *et al.* (1971)
CPFV	*Cucumis sativus*	Sporu	"	Van Dorst & Peters (1974)
HSV	"	Several	"	Sasaki & Shikata (1977a)

[a] Under closely controlled conditions, necrotic local lesions develop on inoculated leaves (Singh, 1973).

temperatures will develop diagnostic symptoms in response to inoculation with mild strains (Diener and Raymer, 1971).

The importance of suitable and rapidly reacting indicator hosts is well illustrated in the search for such hosts for the detection of CEV.

Originally, trifoliate orange was one of few sensitive indicator hosts for the detection of CEV in symptomless citrus trees. This method required 1.5 to more than 5 years from grafting to symptom production (Calavan and Weathers, 1961). Moreira (1961) and Rossetti (1961), by use of shoots and seedlings of Rangpur lime, shortened this period to 4 to 12 months. Later, Calavan et al. (1964) used selections of Etrog citron (Fig. 3) for indexing and were able to obtain a reaction within 1 to 5 months. Finally, the use of Gynura aurantiaca further shortened the incubation period to about 14 days (Weathers and Greer, 1972).

3.3.2 Serological Identification

Before recognition of the unique nature of viroids, it was reasonable for investigators to attempt serological identification procedures in analogy with those used so successfully with conventional plant viruses. Already Chester (1937) included potato spindle tuber among many plant viruses tested for serological reactivity. No precipitation reaction was obtained with PSTV, however. Negative results with PSTV were also reported by Mushin (1942), who performed similar serological tests.

Ball et al. (1964) and Allington et al. (1964), on the other hand, reported the production of high titer antisera specific for PSTV. In microprecipitin and agar diffusion tests, PSTV antiserum reacted with a severe strain of potato virus X and, conversely, PSTV reacted with an antiserum prepared to potato virus X. The authors concluded that PSTV was a strain of potato virus X (Allington et al., 1964). These reports, however, have not been confirmed, and it now seems possible that the plants used for these experiments contained potato virus X as a contaminant (Diener and Raymer, 1971).

Bagnall (1967) prepared antisera that reacted with some component present in the sap of PSTV-infected, but not healthy, tomatoes and potatoes to give a distinctive precipitin line in agar gel diffusion plates. Antiserum titers were low and the "PSTV line" was not seen when either serum or sap was diluted more than four times. Later, however, the antisera were found to contain much material of plant origin and there was no serological evidence of whole virus particles (Singh and Bagnall, 1968). There is no confirmed report of an antiserum to PSTV alone

(Diener and Raymer, 1971). With the recognition that PSTV consists solely of a molecule of RNA, these negative results became readily explicable.

Serological identification, however, again became a possibility when PSTV was recognized as an RNA with a structure similar to that of double-stranded RNA. Immunological tests made with antisera that react specifically with double-stranded RNA (Schwartz and Stollar, 1969), however, gave negative results (Stollar and Diener, 1971), indicating that these antisera do not recognize PSTV as a double-stranded RNA.

No attempts appear to have been made to use serological methods for the identification of any of the other diseases now known to be viroid incited.

3.3.3 Physical-Chemical Identification

Theoretically, a physical and/or chemical test that could discriminate between healthy and viroid-infected tissue should be of considerable use both for practical and research purposes. Such a test could be based either on detection by physical/chemical means of the viroid itself or on detection of an abnormal product occurring in diseased tissue as a consequence of viroid infection. The viroid literature contains one example of each.

3.3.3.1 A Color Test for Exocortis Infection

Childs *et al.* (1958) reported on a color test for CEV infection in *Poncirus trifoliata* trees. The test is based on the observation that ray cells in the bark of infected trees are frequently atypical in unfixed, unstained sections, appearing darker than adjacent cells. The contents of these atypical cells were found to react with aldehyde-decoupling reagents such as phloroglucinol•HCl, giving characteristically colored products that are clearly visible in cross sections of bark under the miscroscope. This reaction is useful as an indication of CEV infection before the onset of bark scaling (Childs *et al.*, 1958). Control experiments showed that the reaction did not occur in unbudded *P. trifoliata* trees of various ages (presumably free of CEV) or in *P. trifoliata* rootstocks known to be free of the viroid. The presence of other pathogens, such as psorosis, xyloporosis, or tristeza, in addition to CEV, did not appear to have an effect on the reaction. In 98.6% of the cases tested, the color reaction correlated with the presence of CEV in *P. trifoliata* infected 2 years or longer.

In Florida, this color test became a part of the indexing program and was used in conjunction with other tests for exocortis in *P. trifoliata* in the Division of Plant Industry (Burnett, 1961). Presumably, with the development of rapid indexing methods (see Section 3.3.1), the color test, which still required 4 years for completion, has become obsolete.

3.3.3.2   Gel Electrophoretic Identification of Potato Spindle Tuber Viroid

Morris and Wright (1975) have developed a diagnostic procedure for the detection of PSTV in small amounts of potato and tomato tissue. The method is a simplified version of a viroid purification scheme (see Chapter 6) and involves extraction of cellular nucleic acids, their separation by polyacrylamide gel electrophoresis, and staining of the nucleic acid bands. Nucleic acid preparations from infected plants are characterized by the appearance in the gel of an extra band, which is due to the presence of PSTV. Both mild and severe strains of PSTV are detected. Use of the test in a routine potato-indexing program showed that it allowed rapid and reliable diagnosis of the disease. This test appears well suited for the elimination of PSTV from elite or basic seed stocks in certification programs (Morris and Smith, 1977).

3.4   BIOASSAY OF VIROIDS

Here we are concerned not with the identification of viroids but with the estimation of viroid concentration by biological means. Other, more accurate methods are available with purified or semipurified viroid preparations (see Chapter 7), but in crude extracts such estimates can only be obtained by measuring the biological activity of viroids in a quantitative, or at least semiquantitative, fashion. Estimation of viroid concentration is essential in efforts to purify and characterize the infectious molecules. Two types of bioassay techniques are used with conventional plant viruses: (1) assay on a host in which the virus spreads systemically, and (2) assay on a hypersensitive host in which the number of local lesions produced is related to the virus concentration used. These two types of bioassay are used with viroids also. Local lesion assays are inherently more accurate than systemic assays. However, the latter type of assay has mostly been used with viroids, because local lesion hosts are not available or because assays based on local lesions posed difficulties.

3.4.1 Assay on Systemic Hosts

The use of systemic infection for the quantitative determination of a biologically active entity is complicated by many factors (see Brakke, 1970). Such assays require large numbers of plants if a reasonably accurate estimate of relative titer is required. Even then, such assays give at best a semiquantitative estimate of the difference between two inocula. Fortunately, much work with viroids does not necessitate an accurate knowledge of relative viroid titer because, in many cases, certain procedures result in large effects and approach all-or-nothing type of responses.

For example, a PSTV preparation that will induce symptoms in essentially all plants when inoculated undiluted or at dilutions of 1/10, 1/100, and 1/1000, and in at least some plants at a dilution of 1/10,000, may be incubated with pancreatic ribonuclease at low concentration, and then inoculated into Rutgers tomato plants. None of the plants will develop symptoms, even if inoculated with the undiluted preparation. Clearly, no statistical analysis is required to ascertain that this result is significant. Similarly, if one studies the elution properties of a viroid from a chromatographic column, for example, or its electrophoretic mobility in polyacrylamide gels, a bioassay based on systemic infection is satisfactory to obtain the desired information.

With experiments of this nature, it is usually sufficient to make a series of 10-fold dilutions of each preparation to be bioassayed and to inoculate each dilution to three to five plants. To obtain accurate results, cross-contamination must be carefully prevented and the assay plants must be inoculated with as uniform a technique as possible.

Several criteria may be used to evaluate results. The least desirable of these is the dilution end point because a large sampling error is associated with the small number of plants infected at the dilution end point (Brakke, 1970). Other criteria are percentage of inoculated plants that become infected at each dilution and time of symptom appearance.

For the assay of PSTV in Rutgers tomato, Raymer and Diener (1969) developed an empirical "infectivity index" that takes into consideration the dilution end point, the percentage of plants infected at each dilution, and the time required for symptom expression. These three factors are used in the index as a means of estimating relative viroid concentration.

As soon as the first plant in an experiment begins to develop symptoms, readings are initiated and are continued at 2-day intervals until no further increase in the number of plants with symptoms has occurred for two or three consecutive readings. The experiment is then terminated. The type of data obtained by this method is illustrated in Table 7, together with the method used to compute an "infectivity index" based on such data.

Evidently, the more concentrated the viroid preparation, the earlier symptoms appear. The index is computed by adding together the number of plants infected at each dilution over the entire recording period, multiplying this figure by the negative log of the dilution, and adding together these products for all the dilutions tested. This index permits discrimination between treatments that might achieve the same dilution end points but differ both in the earliness with which symptoms are expressed and in the number of plants infected. Undoubtedly, the difference in viroid titer is underestimated with this method, since a 10-fold difference in dilution is represented by a difference of only one in the multiplier. On the other hand, the infectivity index gives relatively little weight to individual plants that express symptoms late, after having been inoculated with a highly diluted preparation. Thus much of the variability inherent in an assay of this type is "dampened out."

With careful and consistent inoculation techniques, however, spurious results are rare and in these cases a more realistic index may be computed by using the actual dilution factor as multiplier (Diener, 1971b) instead of the negative logarithm.

Semancik and Weathers (1972a) expressed relative infectivity titers of CEV preparations by determining the total number of "infected plant days" on three to five *Gynura* plants inoculated with one concentration of inoculum only. The authors considered this procedure satisfactory, since the dilution-response curve was believed to be reasonably linear within a certain range of "relative infectivity units" (Semancik and Weathers, 1972a).

Assays similar to the ones described have been used with other viroids. Some details are given in Table 8.

3.4.2 Local Lesion Assay

So far, local lesion hosts have been discovered only for CSV and PSTV. With CSV, local starch lesions (Fig. 12) are detected on inoculated leaves of *Senecio cruentus* (florists' cineraria) 12 to 18 days after inoculation, provided the plants are grown with light intensities not higher than

Table 7. PSTV Symptom Expression in Rutgers Tomato Plants as a Function of Dilution and Time[a] Calculation of an Infectivity Index

Dilution	Days after Inoculation								Sum[c]	Multiplier[d]	Product[e]
	10	12	14	16	18	20	22	24			
10^{-1}	1/3[b]	3/3	3/3	3/3	3/3	3/3	3/3	3/3	22	1	22
10^{-2}				3/3	3/3	3/3	3/3	3/3	15	2	30
10^{-3}					3/3	3/3	3/3	3/3	12	3	36
10^{-4}						1/3	2/3	2/3	5	4	20
10^{-5}							2/3	2/3	4	5	20
10^{-6}							2/3	2/3	4	6	24
10^{-7}							1/3	1/3	2	7	14
10^{-8}								0/3	0	8	0
								Total = Infectivity index:			166

[a] Data were obtained with a viroid concentrate obtained by extraction of infected leaf tissue with 0.5 M K_2HPO_4, chloroform, and n-butanol, followed by phenol treatment and ethanol precipitation. From Raymer and Diener (1969).
[b] Infectivity = number of plants with symptoms/number of plants inoculated.
[c] Sum of all plants showing symptoms at all dates for each dilution.
[d] Negative log of the dilution.
[e] Sum × multiplier for each dilution.

2000 fc and at 18°C to 26°C (Lawson, 1968b.). Increased starch lesion formation is favored by an 18-hour light period with 500 fc fluorescent illumination and a constant temperature of 21°C. Variation in lesion counts among cineraria plants and between leaves on a single plant, however, precludes their use for detecting small quantitative differences in CSV concentration (Lawson, 1968b).

With PSTV, Singh (1971) reported that *Scopolia sinensis*, a solanaceous plant species, produces necrotic local lesions 7 to 10 days after inoculation with the severe strain and 10 to 15 days after inoculation with the mild strain. Singh later stressed that local lesion development is critically dependent on environmental conditions (Singh, 1973). Local lesions developed best on leaves of plants at 22°C to 23°C; plants maintained at 28°C to 31°C developed mostly systemic symptoms without conspicuous local lesions. A low light intensity of 400 fc favored local lesion development.

Figure 12. Lesions induced by CSV in *Senecio cruentus* (florists' cineraria) leaves. Left: Chlorotic spots 35 days after inoculation of left half-leaf. Right: Starch lesions on *S. cruentus* cv. Palette 15 days after inoculation of left half-leaf. Courtesy: R. H. Lawson, U.S. Department of Agriculture, Beltsville, Md.

Table 8. Assay Hosts for Viroids

Viroid	Species	Cultivar	Response	References
PSTV	*Lycopersicon esculentum*	Rutgers	Systemic symptoms	Raymer & Diener (1969)
PSTV	*Scopolia sinensis*	—	Necrotic local lesions	Singh (1971)
CEV	*Gynura aurantiaca*	—	Systemic symptoms	Semancik & Weathers (1972a)
CSV	*Senecio cruentus*	—	Starch lesions	Lawson (1968b)
ChCMV	*Chrysanthemum morifolium*	Deep Ridge	Systemic symptoms	Romaine & Horst (1975)
CPFV	*Lycopersicon esculentum*	Rutgers	Systemic symptoms	Mühlbach & Sänger (1977)
HSV	*Cucumis sativus*	Fukushima	Systemic symptoms	Sasaki & Shikata (1977a)

4. EVIDENCE FOR EXISTENCE OF VIROIDS

Recognition that the agents of viroid diseases have properties drastically different from those of conventional plant viruses came in efforts to purify the putative viruses. This recognition occurred first with PSTV, later with CSV and CEV, and still later with the other known viroid diseases. In this chapter, the crucial evidence that resulted from work with PSTV and led to the viroid concept is presented. Later work with PSTV and the other known viroids is also discussed.

4.1 ORIGINAL WORK WITH POTATO SPINDLE TUBER VIROID

4.1.1 Sedimentation Properties

Diener and Raymer (1967) investigated the sedimentation properties of PSTV in extracts from infected plants. When sap was expressed from infected leaves or when leaves were extracted with phosphate buffer of low ionic strength (0.005 M), most of the infectious material sedimented at low speed together with the tissue debris. Dilution end point of the infectivity from low-speed supernatants was only 10^{-1}. With phosphate buffer of higher ionic strength (0.05 to 0.5 M), however, most of the infectivity was found in the supernatants after low-speed centrifugation. Dilution end points were regularly between 10^{-3} and 10^{-4} (Table 9).

Table 9.   Comparison of Low and High Ionic Strength Buffers for Extraction of PSTV

| Extraction Medium | Infectivity Index[a] | |
	Low-Speed Supernatant	Pellet[b]
0.005 M K$_2$HPO$_4$ + 1% TGA	5	30
0.005 M K$_2$HPO$_4$ + 1% TGA + chloroform-n-butanol[c]	0[d]	72
0.005 M K$_2$HPO$_4$ + 0.1 M ascorbic acid, pH 6.8[e]	7	28
0.5 M K$_2$HPO$_4$ + 1% TGA	48	33
0.5 M K$_2$HPO$_4$ + 1% TGA + chloroform-n-butanol[c]	83[d]	NT

From: Raymer and Diener (1969).

[a] Assayed at dilutions of 10^{-1} to 10^{-4} in 0.005 M K$_2$HPO$_4$.

[b] Low-speed pellet resuspended in original volume of 0.5 M K$_2$HPO$_4$, homogenized for 3 minutes, centrifuged for 15 minutes at 6500 rpm, supernatant diluted and assayed. NT = not tested.

[c] Mixture of chloroform and n-butanol: 1:1 (v/v). TGA = thioglycolic acid.

[d] Aqueous phase.

[e] Extraction medium used by Benson et al. (1964).

4.1.1.1   High-Speed Centrifugation

Efforts to concentrate PSTV by high-speed centrifugation failed (Table 10.) Extracts prepared by grinding infected tissue in 0.5 M K$_2$HPO$_4$ (2 ml per gram of tissue) were clarified by low-speed centrifugation, and the resulting supernatant solutions were then centrifuged at 40,000 rpm (No. 40 rotor, Spinco Model L centrifuge). After centrifugation for 4 hours, most of the infectivity was still in the supernatant solution (dilution end points 10^{-3} to 10^{-4}), and little infectivity was in the resuspended high-speed pellets (10^{-1}).

4.1.1.2   Rate-Zone Centrifugation

To arrive at a more accurate determination of the sedimentation properties of PSTV, extracts were subjected to rate-zonal centrifugation. In preliminary tests, clarified extracts were dialyzed against 0.005 M K$_2$HPO$_4$ and layered onto sucrose density gradients prepared with the same buffer. After centrifugation for 2 to 4 hours at 24,000 rpm (SW 25.1 rotor), the tubes were fractionated and assayed for infectivity. Infectivity was found only near the top of the tubes. Consequently, in further

Table 10. Distribution of PSTV Infectivity before and after High-Speed Centrifugation[a]

Experiment Number	Treatment[b] of Aqueous Phase	Infectivity[c] of						
		High-Speed Supernatant					High-Speed Pellet[d]	
		10^{-1}	10^{-2}	10^{-3}	2×10^{-4}	10^{-4}	10^{-1}	10^{-2}
1[e]	Dialyzed 2 hours vs. 0.005 M K_2HPO_4	3/3	3/3	2/3	NT	0/3	2/3	0/3
2	None (pH 8.2)	3/3	3/3	2/3	1/3	NT	1/3	0/3
2	Adjusted to pH 7.5	3/3	2/3	2/3	1/3	NT	2/3	0/3
2	10^{-4} M Mg^{2+}, pH 8.2	3/3	2/3	2/3	2/3	NT	1/3	0/3
2	10^{-3} M Mg^{2+}, pH 8.2	3/3	2/3	1/3	2/3	NT	1/3	0/3

From: Raymer and Diener (1969).

[a] Centrifuged 1 hour at 40,000 rpm, No. 40 rotor, Spinco Model L ultracentrifuge.

[b] Tissue extracted with 0.5 M K_2HPO_4, chloroform, and n-butanol (1:2:1:1, w:v:v:v). Phases separated by centrifugation for 15 minutes at 6000 to 6500 rpm; aqueous phase used for tests.

[c] Number of plants infected/number of plants inoculated. NT = not tested. All preparations were diluted for assay in 0.005 M K_2HPO_4.

[d] Resuspended in original volume of 0.5 M K_2HPO_4. No infectivity at dilutions of 10^{-3}, 2×10^{-4}, or 10^{-4}.

[e] The aqueous phase of experiment 1 yielded the following results on bioassay: 3/3 plants infected at a dilution of 10^{-1}, 2/3 at 10^{-2}, 2/3 at 10^{-3}, and 2/3 at 10^{-4}.

tests, centrifugation was extended to 16 hours. Results of experiments with high-speed supernatants and with resuspended high-speed pellets are shown in Table 11.

Regardless of whether organic solvents were used during the extraction of the tissue, infectivity of high-speed supernatants was confined to the upper and middle portions of the gradient columns. Distribution of infectivity from resuspended high-speed pellets was similar, but as expected, levels of infectivity were lower than in the high-speed supernatants. The high-speed pellets from extracts prepared without organic solvents contained considerable amounts of green material. This material was large enough to sediment to the bottom of the tubes during density-gradient centrifugation. Considerable infectivity was associated with these pelleted fractions (Table 11).

Absorbance profiles of centrifuged sucrose gradients that contained clarified extracts (or high-speed supernatants) from healthy or PSTV-infected tissue did not disclose qualitative differences or well-defined components. Better absorbance profiles were achieved with ethanol-concentrated preparations, and since such preparations had very high infectivity dilution end points (up to 10^{-7}), numerous correlations between location of infectivity within a gradient and the absorbance profile of the gradient were attempted (Diener and Raymer, 1969).

Figure 13 shows the absorbance profiles of ethanol concentrates prepared in identical fashion from equal amounts of healthy and PSTV-infected tissue and infectivity indices of successive 2-ml fractions collected from the latter gradient. Evidently, no component with sedimentation properties identical with the infectious entity and present only in the extract from infected leaves could be found. Disregarding minor variations, the two absorbance profiles were identical.

The slow rate of sedimentation of the infectious material made it unlikely that the extracted infectious agent was a conventional viral nucleoprotein. It appeared more likely that this material was a free nucleic acid. Consequently, treatments customarily used for nucleic acid extraction and concentration were applied to PSTV extracts.

PSTV-infected tissue was homogenized in 0.5 M K_2HPO_4 alone or together with a mixture of chloroform and butanol. Aliquots of the clarified extracts were then treated three times in succession with equal volumes of phenol. Residual phenol was removed from the aqueous phase by repeated ether extraction. After appropriate dilution, the preparations were bioassayed. In other experiments, tissue was extracted

Table 11.   Distribution of Infectivity in Centrifuged Sucrose Density Gradients[a] of PSTV Extracts

Fraction[b] Number	Expressed Sap[d]	Without Organic Solvents[e]		With Organic Solvents[f]	
		HS[g]	HP[h]	HS	HP
1	0/3	0/3	0/3	0/3	0/3
2	0/3	3/3	2/3	0/3	1/3
3	0/3	3/3	0/3	2/3	1/3
4	0/3	2/3	0/3	3/3	0/3
5	0/3	3/3	0/3	0/3	0/3
6	0/3	2/3	0/3	0/3	0/3
7	0/3	2/3	1/3	3/3	0/3
8	0/3	0/3	0/3	1/3	0/3
9	0/3	0/3	0/3	0/3	0/3
10	0/3	0/3	0/3	0/3	0/3
11	0/3	0/3	0/3	0/3	0/3
12	1/3	0/3	0/3	0/3	0/3
Resuspended pellet	3/3	0/3	3/3	0/3	0/3

From Raymer and Diener (1969).

[a] Linear gradients, 0.2 to 0.7 M sucrose in 0.005 M K_2HPO_4, centrifugation for 16 hours at 24,000 rpm in Spinco L ultracentrifuge, SW 25.1 rotor.

[b] Consecutive 2-ml fractions. Fraction 1 is from top, fraction 12 from bottom of tube. All fractions diluted 1:4 with 0.005 M K_2HPO_4 for inoculation. Pellets resuspended in 2 ml of same buffer for inoculation.

[c] Number of plants with symptoms/number of plants inoculated.

[d] No buffer, solvent, clarification, or high-speed centrifugation used prior to density-gradient centrifugation.

[e] Tissue extracted with 0.5 M K_2HPO_4 (2 ml/g tissue), extract clarified at 12,000 rpm for 15 minutes, supernatant centrifuged 1 hour at 40,000 rpm.

[f] Tissue extracted with 0.5 M K_2HPO_4 + chloroform + n-butanol (2:1:1 ml/g of tissue), centrifuged for 15 minutes at 6500 rpm. Aqueous phase centrifuged for 1 hour at 40,000 rpm.

[g] HS = high-speed supernatant.

[h] HP = high-speed pellet resuspended in original volume of 0.005 M K_2HPO_4.

directly with 0.5 M K_2HPO_4 and phenol. After centrifugation and removal of residual phenol the aqueous phases were bioassayed. As shown in Table 12, treatment of extracts with phenol did not lead to an appreciable change of the infectivity levels of the preparations (Raymer and Diener, 1969).

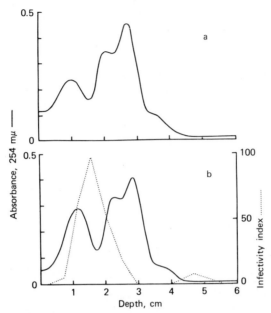

Figure 13. Absorbance profiles of centrifuged density-gradient columns containing ethanol-concentrated extracts from healthy and PSTV-infected tissue, and infectivity distribution in the gradient containing the extract from infected tissue. (*a*) Extract from healthy tissue; 1 ml of a solution with 3.5 OD_{260} units/ml was layered onto the gradient. (*b*) Extract from infected tissue; 1 ml of a solution with 3.5 OD_{260} units/ml was layered onto the gradient. (———) = absorbance profile; (• • •) = infectivity indices of consecutive 2-ml fractions collected from gradient. From: Diener and Raymer (1969).

To investigate the effects of treatment with phenol on the sedimentation properties of PSTV, extracts were subjected to density-gradient centrifugation before and after treatment with phenol. As the sedimentation behavior of single-stranded nucleic acids is known to depend on the ionic strength of the suspending medium, analyses were performed in gradients with both low and high ionic strength. An extract from PSTV-infected tissue (0.5 *M* K_2HPO_4-chloroform-butanol) was subjected to high-speed centrifugation. The supernatant solution was then divided into two portions. One portion was dialyzed versus 0.003 *M* K_2HPO_4 and was analyzed in a sucrose gradient containing 0.005 *M* K_2HPO_4. A sample of the other portion was directly analyzed in a sucrose gradient containing 0.5 *M* K_2HPO_4. Another sample was concentrated by ethanol precipitation, resuspended in 0.5 *M* K_2HPO_4, and analyzed in a gradient containing 0.5 *M* K_2HPO_4. A sample of the concentrated extract was treated with phenol and, after removal of residual

Table 12. Effect of Phenol Treatment on Infectivity of PSTV

Method of Extraction	Fraction Assayed	Infectivity[a] of							
		Extract, Diluted				Phenol-Treated Extract, Diluted			
		10^{-1}	10^{-2}	10^{-3}	10^{-4}	10^{-1}	10^{-2}	10^{-3}	10^{-4}
$0.5\ M$ K$_2$HPO$_4$	Clarified extract	NT	3/3	1/3	0/3	NT	3/3	1/3	0/3
	High-speed supernatant	NT	3/3	2/3	0/3	NT	3/3	3/3	1/3
	High-speed pellet	NT	0/3	0/3	0/3	NT	0/3	0/3	0/3
$0.5\ M$ K$_2$HPO$_4$ + CHCl$_3$ + BuOH	Clarified extract	NT	3/3	2/3	0/3	NT	3/3	1/3	0/3
	High-speed supernatant	NT	3/3	0/3	0/3	NT	3/3	2/3	0/3
	High-speed pellet	NT	3/3	NT	0/3	NT	1/3	0/3	0/3
$0.5\ M$ K$_2$HPO$_4$	Aqueous phase[b]	NT	NT	NT	NT	3/3	3/3	1/3	0/3
$0.5\ M$ K$_2$HPO$_4$ + CHCl$_3$ + BuOH	Clarified extract	3/3	3/3	2/3	0/3	3/3	2/3	1/2	0/3

From: Raymer and Diener (1969).

[a] Number of plants with symptoms/number of plants inoculated and surviving. Inocula diluted in 0.005 M K$_2$HPO$_4$. NT = not tested.

[b] Tissue extracted directly with 0.5 M K$_2$HPO$_4$ and phenol.

phenol, the aqueous phase was analyzed in a gradient containing 0.5 M K_2HPO_4. After centrifugation, the four tubes were fractionated by collecting consecutive 2-ml fractions from each tube. All fractions were assayed for infectivity after dilution with 0.005 M K_2HPO_4. Table 13 shows the results.

Although centrifugation of the supernatant solution in 0.5 M K_2HPO_4 resulted in a somewhat wider distribution of infectivity within the gradient than in 0.005 M K_2HPO_4, most of the infectivity in either case was contained in the top four to five fractions of the gradient. In the tubes containing the concentrate, infectivity spread was much wider than in the tubes containing the supernatant solution, whether analysis was performed at low or high ionic strength. Treatment of concentrates with phenol resulted in infectivity distributions similar to those found with concentrates that had not been treated with phenol; that is most of the infectivity was contained in the upper and middle fractions of the gradients (Diener and Raymer, 1969).

Figure 14 shows the absorbance profiles and infectivity indices of a phenol-treated concentrate prepared from PSTV-infected tissue. Similar extracts from healthy leaves had essentially identical absorption profiles. Phenol-treated extracts had ultraviolet spectra typical of nucleic acids

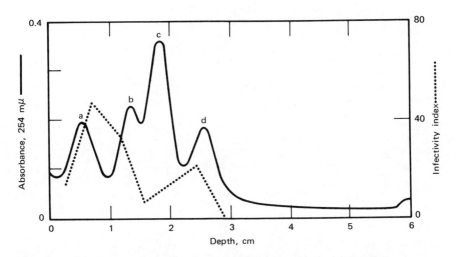

Figure 14. Absorbance profile and infectivity indices of a centrifuged density-gradient column containing a phenol-treated concentrate from PSTV-infected leaves. One ml of a solution with 2.5 OD units/ml was layered onto the gradient. a = tRNA, b = DNA, c and d = rRNAs. From: Diener and Raymer (1969).

Table 13. **Effect of Ionic Strength and of Treatment with Phenol on the Sedimentation Properties of PSTV in Sucrose Density Gradients**

Sample Assayed[a]	Molar Concentration of K_2HPO_4 in Gradient	Infectivity[b] of Consecutive 2-ml Fractions from Centrifuged Gradient Tubes[c]												
		1	2	3	4	5	6	7	8	9	10	11	12	P[d]
High-speed supernatant	0.005[e]	0/3	1/3	2/3	1/3	1/3	0/3	0/3	0/3	0/3	0/3	0/3	0/3	0/3
High-speed supernatant	0.5	3/3	3/3	3/3	2/3	2/3	1/3	1/3	1/3	0/3	0/3	0/3	0/3	1/3
Concentrate	0.005	3/3	3/3	3/3	2/3	2/3	2/3	1/3	1/3	1/3	1/3	1/3	0/3	0/3
Concentrate	0.5	3/3	3/3	3/3	3/3	3/3	0/3	3/3	3/3	3/3	1/3	2/3	1/3	3/3
Phenol-treated	0.005	0/3	1/3	3/3	3/3	3/3	2/3	1/3	3/3	0/3	1/3	1/3	1/3	0/3
concentrate	0.5	3/3	3/3	2/3	1/3	2/3	1/3	1/3	1/3	1/3	1/3	0/3	0/3	1/3

From: Diener and Raymer (1969).

[a] PSTV-infected tissue was extracted with 0.5 M K_2HPO_4, chloroform, and butanol; the aqueous phase was used for the experiment.

[b] Number of plants with symptoms/number of plants inoculated.

[c] Fraction 1 is from top of gradient; fraction 12 is from bottom of gradient. Fractions diluted 1/10 in 0.005 M K_2HPO_4.

[d] P = Pellet resuspended in 2 ml of 0.005 M K_2HPO_4.

[e] Sample dialyzed vs. 0.003 M K_2HPO_4 before analysis. Low infectivity was presumably due to losses incurred during dialysis.

with maxima at 258 nm, minima at 230 nm, and maximum–minimum ratios of 2.3 to 2.4. Incubation of these extracts with RNase or DNase, followed by density-gradient centrifugation, disclosed that peaks *a, c,* and *d* were RNA and that peak *b* was DNA (Fig. 14). Comparison of these profiles with profiles reported with nucleic acid preparations from plant tissues (Ralph and Bellamy, 1964) made it evident that peak *a* corresponded to transfer RNA (tRNA), and peaks *c* and *d* corresponded to the two species of ribosomal RNA (rRNA). The major portion of PSTV sedimented in sucrose gradients with approximately 10 S, based on rRNAs (17 and 27 S) used as markers.

These experiments clearly showed that the infectious material extractable from PSTV-infected tissue with phosphate buffer has a very low sedimentation coefficient and that this property is not significantly affected by subsequent treatment of the preparations with phenol, suggesting that the material originally extracted already was a free nucleic acid (Diener and Raymer, 1967, 1969).

4.1.1.3 Equilibrium Centrifugation

Preliminary experiments had shown that exposure of PSTV extracts to cesium chloride concentrations as high as nine molal for 3 days at 4°C did not lead to an appreciable reduction in the infectivity of the preparations.

Concentrates from PSTV-infected tissue (not treated with phenol) were brought to a density of 1.70 g/cm³ with cesium chloride or to a density of 1.617 g/cm³ with cesium sulfate. After centrifugation, the tubes were fractionated and consecutive 0.5-ml fractions were collected, diluted, and bioassayed.

Figure 15 shows the results. The cesium chloride gradient contained, as expected, only one ultraviolet-absorbing peak, namely the DNA present in the sample, whereas all RNA had pelleted. All infectivity had similarly pelleted during centrifugation (Fig. 15a). The cesium sulfate gradient contained one major ultraviolet-absorbing peak with pronounced shoulders on either side of this peak. Infectivity roughly coincided with the major peak, but a small amount of infectivity was also recovered from the pellet (Fig. 15b). These observations are in accord with the view that the infectious material is RNA, not DNA (Diener and Raymer, 1969).

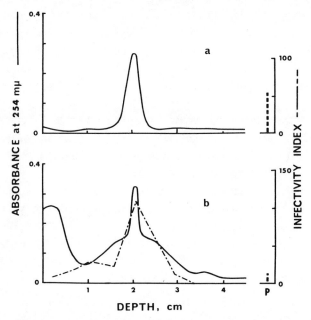

Figure 15. Absorbance profiles and infectivity distributions in cesium chloride (*a*) and cesium sulfate (*b*) gradients that contained PSTV concentrate (not treated with phenol) and were centrifuged for 72 hours at 35,000 rpm. P = resuspended pellet. From: Diener and Raymer (1969).

4.1.2   Sensitivity to Nucleases

In view of these results, it was of interest to investigate whether the infectivity in extracts from PSTV-infected leaves was sensitive to treatment with nucleases.

4.1.2.1   Endonucleases

Aliquots of phenol-treated concentrates suspended in 0.005 M K$_2$HPO$_4$ (unadjusted, pH 8.2) were incubated in the presence of pancreatic ribonuclease (RNase) or deoxyribonuclease (DNase). Control aliquots were incubated in the absence of nucleases. After incubation, the preparations were rapidly cooled to 0°C, diluted, and bioassayed. Other aliquots, to which equal amounts of RNase or DNase had been added, but which were not incubated at 25°C, were similarly diluted and bioassayed.

Table 14 shows that no loss of infectivity occurred when the extract was incubated in the absence of nucleases. Addition of RNase to the extract led to a drastic reduction in infectivity even without incubation. After incubation with RNase for 1 hour at 25°C, no infectivity remained. Treatment with DNase, on the other hand, had no effect on the infectivity of PSTV. Analyses of the untreated and DNase-treated extracts by density-gradient centrifugation disclosed that the DNA peak present in the untreated extract had completely disappeared after incubation with DNase (Diener and Raymer, 1967, 1969).

In other experiments, the question was asked whether quantitative differences existed in the RNase sensitivity of phenol-treated extracts as compared with those not treated with phenol. In this case, high-speed supernatants were used instead of phenol-treated concentrates. Thus the viroid was suspended in 0.5 M K_2HPO_4. Two levels of RNase concentration were used. After incubation, the preparations were diluted with 0.005 M K_2HPO_4 and infectivities of extracts not incubated or incubated for 2 hours at 25°C were compared. Table 15 shows that no appreciable difference was found in the sensitivity of the two extracts to RNase. It can be seen, however, that some infectivity survived all RNase treatments in the undiluted incubation mixtures.

The question thus arose whether PSTV was protected from RNase attack when suspended in media of high ionic strength. In one experiment, RNase-induced loss of infectivity in 0.15 M NaCl—0.015 M sodium citrate, pH 7 (SSC) was compared with that in 0.005 M K_2HPO_4 (Table 16), and in another experiment, RNase sensitivities in neutral

Table 14. Sensitivity of PSTV to Treatment with Nucleases

Enzymatic Treatment	Incubation at 25°C (hours)	Infectivity Index[a]
None	0	150
None	1	183
1 μg/ml RNase	0	12
1 μg/ml RNase	1	0
1 μg/ml DNase[b]	0	156
1 μg/ml DNase[b]	1	171

From: Diener and Raymer (1969).

[a] See Chapter 3. All preparations were assayed at dilutions of 10^{-1}, 10^{-2}, 10^{-3}, 10^{-4}, and 10^{-5}, made in 0.005 M K_2HPO_4.

[b] Incubation medium contained 10^{-2} M $MgCl_2$.

Table 15. Sensitivity of Clarified Extracts and of Phenol-Treated Extracts from PSTV-Infected Tissues to RNase[a]

Enzymatic Treatment	Incubation at 25°C (hours)	Infectivity[b] of							
		Extract, Diluted				Phenol-Treated Extract, Diluted			
		None	10^{-1}	10^{-2}	10^{-3}	None	10^{-1}	10^{-2}	10^{-3}
None	0	3/3	3/3	3/3	3/3	3/3	3/3	2/3	2/3
None	2	3/3	3/3	3/3	2/3	3/3	3/3	3/3	2/3
0.1 µg/ml RNase	0	3/3	3/3	3/3	0/3	3/3	3/3	3/3	2/3
0.1 µg/ml RNase	2	3/3	3/3	3/3	0/3	3/3	0/3	0/3	0/3
1 µg/ml RNase	0	3/3	0/3	0/3	0/3	3/3	0/3	0/3	0/3
1 µg/ml RNase	2	3/3	0/3	0/3	0/3	3/3	0/3	0/3	0/3

From: Diener and Raymer (1969).

[a] PSTV-infected tissue was extracted with 0.5 M K_2HPO_4, chloroform, and butanol; the aqueous phase was centrifuged for 1 hour at 40,000 rpm. The resulting supernatant solution was used for the experiment.

[b] Number of plants with symptoms/number of plants inoculated. Dilutions were made with 0.005 M K_2HPO_4.

Table 16. Effect of Ionic Strength on RNase Sensitivity of Phenol-Treated Concentrates of PSTV[a]

Suspending Medium	RNase Concentration	Infectivity Index[b]	
		Not Incubated	Incubated 1 Hour at 25°C
0.005 M K$_2$HPO$_4$	0	NT	185
	0.01	187	16
	0.1	11	0
SSC[c]	0	NT	216
	0.01	193	121
	0.1	192	12

From: Diener and Raymer (1969).

[a] PSTV-infected tissue was extracted with 0.5 M K$_2$HPO$_4$, chloroform, and butanol; the aqueous phase was separated, treated with phenol, and concentrated by ethanol precipitation. The pellets were resuspended in the respective buffer and used for the experiment.

[b] All preparations were assayed at dilutions of 10^{-1}, 10^{-2}, 10^{-3}, 10^{-4}, and 10^{-5}, made with 0.005 M K$_2$HPO$_4$. NT = not tested.

[c] SSC = 0.15 M NaCl—0.015M sodium citrate, pH 7.

phosphate buffers of low and high ionic strengths were determined (Table 17).

It can be seen that in SSC infectivity partially survived the RNase treatment, and that a relatively large proportion of infectivity survived the RNase treatment when the samples were incubated in 1 M phosphate buffer, pH 7.

It is known, however, that single-stranded RNAs are also more stable to treatment with RNase in media of high ionic strength (Billeter *et al.*, 1966). To investigate the effect of high ionic strength on the inactivation by RNase of a single-stranded viral RNA, experiments were made with southern bean mosaic virus (SBMV)-RNA. SBMV-RNA was precipitated with ethanol and dissolved in phenol-treated concentrates prepared from PSTV-infected tissue that had previously been dialyzed versus 0.005 M or 1.0 M K$_2$HPO$_4$. SBMV-RNA concentration was 0.1 mg/ml in each preparation. Thus the total nucleic acid concentration was equal to or higher than that in solutions used for the investigation of RNase sensitivity of PSTV, and equal or higher competition of nonviral RNA for the enzyme was present. RNase was then added to aliquots of both preparations. These samples, together with RNase-free controls, were

Table 17.   Effect of Ionic Strength of Neutral Phosphate Buffers on RNase Sensitivity of PSTV Concentrate[a]

RNase Concentration (μg/ml)[b]	Diluent, Phosphate Buffer, pH 7 (M)	Infectivity Index[c] of Sample Incubated in	
		0.005 M Phosphate, pH 7	1.0 M Phosphate, pH 7
0	0.005	78	69
0.1	0.005	0	69
0.5	0.005	0	53
1.0	0.005	0	22
0	1.0	NT	35
0.1	1.0	NT	55
0.5	1.0	NT	42
1.0	1.0	NT	33

From: Diener and Raymer (1969).
[a] A concentrate of PSTV was dialyzed vs. 0.005 M or 1.0 M phosphate buffer, pH 7, and used for the experiment.
[b] Incubation for 1 hour at 25°C.
[c] See Chapter 3. All preparations were assayed undiluted and at dilutions of 10^{-1}, 10^{-2}, 10^{-3}, 10^{-4}, and 10^{-5}. NT = not tested.

incubated for 1 hour at 25°C. After dilution in the appropriate buffer, the preparations were assayed for infectivity.

Incubation with RNase completely inactivated SBMV-RNA under the conditions used, whether the RNA was dissolved in a medium of low or high ionic strength (data not shown). Thus the resistance of PSTV to RNase could not be explained solely by inhibition of RNase activity or by increased resistance of single-stranded RNA under conditions of high ionic strength.

The observed sensitivity of PSTV to RNase (at low ionic strength) and its insensitivity to DNase indicated that the infectious molecule is an RNA or that at least one portion, which is essential for infectivity, is composed of RNA (Diener and Raymer, 1967, 1969). Also, the partial resistance of PSTV to RNase under conditions of high ionic strength suggested that the RNA might be double stranded.

4.1.2.2   Exonucleases

To determine whether PSTV was sensitive to inactivation by exonucleases, phenol-treated preparations of PSTV were incubated with snake venom phosphodiesterase. As an internal control, TMV RNA was added to the preparations. Since tomato is a host for both TMV and PSTV, and since symptoms are readily distinguishable, preparations could be assayed for TMV and PSTV on one set of plants. TMV symptoms appeared within 1 week after inoculation, whereas those of PSTV appeared after 2 to 3 weeks. Presence or absence of TMV was verified by transfer to Pinto bean plants, a local lesion host of TMV. In all cases, extracts from tomato plants without TMV symptoms incited no lesions on Pinto, and extracts from tomato plants with TMV symptoms incited abundant lesions on Pinto (Diener, 1970, 1971c).

As shown in Table 18 (experiment 1), incubation with snake venom phosphodiesterase completely inactivated TMV RNA, yet had no effect on the infectivity of PSTV. These results indicate that PSTV is either circular or is "masked" at the 3'-terminus in such a fashion that the

Table 18. Effect of Incubation with Snake Venom Phosphodiesterase and Alkaline Phosphatase on the Infectivity of PSTV and TMV RNA

Experiment Number	Treatment[a]	Infectivity Index[b]	
		PSTV	TMV RNA
1	Control, no incubation	136	66
	Control, 1 hour at 25°C	164	69
	Phosphodiesterase added after incubation	156	60
	1 hour at 25°C with phosphodiesterase	165	0
2	Control, 1 hour at 25°C	138	240
	1 hour at 25°C with alkaline phosphatase	144	186
	1 hour at 25°C with phosphodiesterase	185	0
	1 hour at 25°C with alkaline phosphatase and phosphodiesterase	185	0

From: Diener (1971c).

[a] Phenol-treated preparations of PSTV were mixed with TMV RNA and dialyzed against 0.02 M glycine—NaOH, 3 mM $MgCl_2$, pH 9.0. Final concentration of snake venom phosphodiesterase (Worthington, *Crotalus adamanteus*), 166 μg/ml; of alkaline phosphatase (Worthington, *E. coli*, code BAPF), 33 μg/ml.

[b] Assayed at dilutions of 10^{-1}, 10^{-2}, 10^{-3}, and 10^{-4}. In experiment 1, TMV RNA and PSTV were mixed and assayed on one set of tomato plants; in experiment 2, PSTV and TMV RNA were treated and assayed separately.

enzyme cannot attack the terminal nucleotide. It is well known that nucleic acids phosphorylated at the 3'-terminus (the 5'-linked end) are resistant to snake venom phosphodiesterase (Singer and Fraenkel-Conrat, 1963). Consequently, in further experiments PSTV was incubated with a mixture of snake venom phosphodiesterase and alkaline phosphatase (Table 18, experiment 2). Enzymic activity of the alkaline phosphatase used was verified by colorimetric assay of the enzyme. Evidently, PSTV resisted attack by a combination of exonuclease and alkaline phosphatase. Thus resistance of PSTV to snake venom phosphodiesterase could not be explained by a phosphorylated 3'-terminus.

It was of interest to determine whether PSTV was susceptible to inactivation by exonucleases that attack the nucleic acid from the 5'-terminus. As shown in Table 19, incubation of PSTV with bovine spleen phosphodiesterase alone, or in combination with alkaline phosphatase, had no effect on the infectivity of PSTV (Diener, 1970, 1971c).

On the basis of these results, the hypothesis was advanced that PSTV might be a circular RNA molecule (Diener, 1970).

4.1.3   Absence of Virus Particles

Although there was little doubt that the slowly sedimenting infectious entities in preparations from PSTV-infected tissues were free RNA molecules, the question arose as to whether these infectious entities existed

Table 19.   Effect of Incubation with Bovine Spleen Phosphodiesterase on the Infectivity of PSTV

Treatment[a]	Infectivity Index[b]
Control, 1 hour at 25°C	124
Phosphodiesterase added after incubation	99
1 hour at 25°C with phosphodiesterase	84
1 hour at 25°C with phosphodiesterase and alkaline phosphatase	94

From: Diener (1971c).

[a] Bovine spleen phosphodiesterase (Schwarz) and alkaline phosphatase (Worthington, *E. coli,* code BAPF) were added to portions of a phenol-treated preparation of PSTV in 0.02 M phosphate buffer, pH 8.0. Final concentration of bovine spleen phosphodiesterase, 0.33 units/ml; of alkaline phosphatase, 166 μg/ml.

[b] Assayed at dilutions of 10^{-1}, 10^{-2}, 10^{-3}, and 10^{-4}.

as such *in situ* or whether they were released from conventional virus particles during extraction.

A systematic study of this question led to results that were incompatible with the concept that conventional viral nucleoprotein particles exist in PSTV-infected tissue (Diener, 1971a).

Examination of Figure 13 reveals that in addition to the slowly sedimenting infectious material, small quantities of faster-sedimenting infectious material are also present in PSTV-containing extracts. Conceivably, this material could consist of viral nucleoprotein particles. If one assumes that the free infectious RNA that was present in all extracts was derived from virions that had been degraded during extraction, the observed heterogeneity of the infectious material could be explained by the presence of complete virions, as well as of virions in various stages of degradation. In this case, one would expect the faster-sedimenting infectious material to be more stable during storage, and more resistant to attack by RNase, than the liberated RNA. This, however, was not the case. On the contrary, as shown in Table 20, the longevity of the faster-sedimenting material was less than that of the more slowly sedimenting material. Also, incubation of extracts with RNase led to inactivation of all infectivity, including that associated with the faster-sedimenting material (Diener and Raymer, 1969). Furthermore, neither incubation of extracts with Pronase nor treatment of extracts with phenol or with phenol in the presence of sodium dodecyl sulphate (SDS) significantly affected the sedimentation properties of the infectious particles (Table 21) (Diener, 1971a).

These results make it unlikely that the RNA is bound to protein or that it is present in the form of complete or partially degraded virions. This conclusion was strengthened by the observation that highly purified RNA preparations from PSTV-infected tissue also contain infectious material that sediments at faster rates than would be expected of free RNA (Diener, 1971a). If the faster-sedimenting infectious material were composed of viral nucleoprotein particles, these particles would have most unusual properties. On the one hand, their protein coat would have to be loose enough to allow access to RNase (since all infectivity is sensitive to treatment with RNase); yet, on the other hand, these structures would have to be resistant to treatment with phenol and SDS. As there are apparently no known nucleoproteins with these properties, Diener (1971a) concluded that the faster-sedimenting infectious material was not composed of viral nucleoprotein particles.

Table 20. Longevity of Infectious Material in Extracts[a] from PSTV-Infected Tissue

	Infectivity Indices		
Fraction[b] Number	Immediately	After 1 Day at 4°C	After 7 Days at 4°C
1	17	18	14
2	19	11	13
3	19	12	3
4	16	12	1
5	15	13	1
6	14	15	6
7	16	12	0
8	10	4	0
9	12	2	0
10	16	2	0
11	15	0	0
12	14	2	0
13	10	0	0
P	15	13	0

From: Diener (1971a).

[a] Five grams of fresh PSTV-infected tissue were ground at 4°C with 20 ml of 0.05 M K_2HPO_4 (mortar and pestle); the extract was adjusted to pH 8.3 with dilute NaOH and was then centrifuged for 15 minutes at 5000 g. The supernatant was analyzed.

[b] Consecutive 2-ml fractions. Fraction 1 is from the top; fraction 13 is from the bottom of the tube. P = Pellet resuspended in 0.02 M phosphate buffer, pH 7. Linear sucrose gradients, 0.2 to 0.8 M in 0.02 M phosphate buffer, pH 7; centrifugation for 2.5 hours at 24,000 rpm; SW 25.1 rotor, Spinco Model L centrifuge.

This conclusion was further strengthened by the observation that, *in situ*, PSTV is sensitive to treatment with RNase. Vacuum infiltration of RNase solutions into PSTV-infected leaves resulted in complete or nearly complete loss of infectivity under conditions where the infectivity of conventional virus particles was not affected (Diener, 1971c).

In Table 22 the protocol for these experiments is shown and in Table 23 some of the results are given.

Zaitlin and Hariharasubramanian (1972) analyzed proteins isolated from PSTV-infected leaves and compared them with proteins isolated from uninfected leaves. They found no evidence for the production in infected leaves of proteins that could be construed as coat proteins

Table 21. Effect of Treatment with Phenol and with Phenol-SDS on Infectivity Distribution in Sucrose Gradients of Crude PSTV Extracts[a]

Fraction[b] Number	Infectivity Indices		
	Control	Phenol	Phenol-SDS
1	6	15	13
2	14	15	17
3	11	16	13
4	11	14	7
5	8	16	8
6	13	7	0
7	6	9	0
8	6	9	7
9	9	5	9
10	7	0	15
11	7	0	9
12	11	4	9
13	10	5	NT
P	15	2	5

From: Diener (1971a).

[a] Five grams of fresh PSTV-infected tissue were ground with 20 ml of 0.01 M K_2HPO_4 (mortar and pestle); the extract was adjusted to pH 8.3 with dilute NaOH and was then centrifuged for 15 minutes at 5000 g. The resulting supernatants were used. One portion was kept as a control. To another portion, 1 volume of water-saturated phenol was added in one experiment. In the other experiment, enough solid SDS was added to give an SDS concentration of 1.5%. One volume of phenol was then added. The resulting emulsions were vigorously shaken by hand for 5 minutes at room temperature and were then centrifuged to separate the aqueous from the phenol phases. The aqueous phases were then shaken three times in succession with equal volumes of ether, the ether phases being removed each time; residual ether was removed by bubbling of nitrogen through the solution. The resulting solutions were analyzed.

[b] Consecutive 2-ml fractions. Fraction 1 is from the top; fraction 13 is from the bottom of the tubes. P = Pellet resuspended in 0.02 M phosphate buffer, pH 7. Centrifugation for 2.5 hours at 24,000 rpm, SW 25.1 rotor.

under conditions where coat protein of a defective strain of tobacco mosaic virus could readily be demonstrated.

Electron microscopy of thin sections prepared from PSTV-infected and healthy tissue failed to reveal viruslike particles in cells of infected plants (R. H. Lawson, personal communication).

In light of these results, Diener (1971a) concluded that it appeared

Table 22.   Protocol for Nuclease Infiltration Experiments

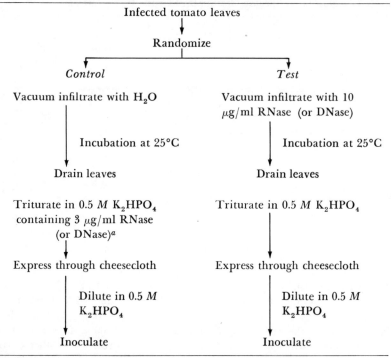

From: Diener (1971a).

[a] Amount of nuclease added based on assumption of equilibration of nuclease within leaves of test sample.

unlikely that virions existed in PSTV-infected tissue and that, even if such structures should exist, they would have to be extremely fragile and could not play a significant role in the natural transmission of the disease agent.

4.1.4   Recognition of Low Molecular Weight

The low sedimentation rate of PSTV is consistent with a viral genome of conventional size ($\geq 10^6$ daltons) only if the RNA is double-stranded or multistranded. Early experiments indeed indicated that the RNA might be double-stranded (Diener and Raymer, 1969), but later results showed that its chromatographic properties are not compatible with this hypothesis (Diener, 1971c) (see Chapter 7).

Table 23. Effect of Infiltration[a] of PSTV-Infected Tomato Leaves with Ribonuclease or Deoxyribonuclease on Infectivity

Experiment Number	Treatment	Time of Incubation (hours at 25°C)	Infectivity Index[b]
1	RNase-infiltrated	1	62
	Water-infiltrated[c]	1	68
2	RNase-infiltrated	22	0
	Water-infiltrated[c]	22	98
3	RNase-infiltrated[d]	25	24
	Water-infiltrated[d]	25	87
4	RNase-infiltrated	24	4
	Water-infiltrated	24	55
5	DNase-infiltrated	22	63
	Water-infiltrated	22	76

From: Diener (1971a).

[a] See Table 22 for procedure.

[b] Preparations assayed undiluted and at dilutions of 10^{-1}, 10^{-2}, 10^{-3}, and 10^{-4}.

[c] No RNase added to control.

[d] Bentonite added to grinding buffer.

Evidently, determination of the molecular weights of viroids was of great importance in elucidating their structure. This determination was difficult because the agents occur in infected tissue in very small amounts and are therefore difficult to separate from host RNA and to purify in amounts sufficient for conventional biophysical analyses.

A principle elaborated by Loening (1967), however, made it possible to obtain a first estimate of the molecular weight of PSTV based solely on biological activity. Loening reasoned that the effects of secondary structure of an RNA on its sedimentation properties are opposite to the effects of secondary structure on its electrophoretic mobility in polyacrylamide gels. Unfolding of an RNA molecule should lower its electrophoretic mobility so that it will move with molecules of higher molecular weight in polyacrylamide gels, whereas it will move with molecules of lower molecular weight during sedimentation. Thus a combination of the two analytical methods should permit distinguishing differences in structure from differences in weights of RNAs. With either method, biological activity is the only parameter necessary for evaluation of the results. Application of this principle to PSTV led to the following unexpected results.

An analysis of the sedimentation properties of PSTV in highly purified nucleic acid preparations from infected tissues, together with those of nucleic acid markers of known sedimentation coefficients, is illustrated in Figure 16. Peak infectivity was associated with the fraction that corresponds to an $S_{20,w}$ value of 5.5. A minor peak of infectivity was found at $S = 7.0$ (Diener, 1971b).

Several authors have proposed empirical formulae that relate sedimentation coefficients of RNAs to their molecular weights. The application of one of these, a formula proposed by Gierer (1958), to PSTV results in a molecular weight estimate of 4.7×10^4 for the major fraction and 7.5×10^4 for the minor fraction (Diener, 1971b). These values, however, are unreliable so long as the secondary and tertiary structure of PSTV is not known.

Analysis of PSTV, together with marker RNAs, in 3% polyacrylamide gels revealed two major infectivity peaks (Fig. 17). Highest infectivity was in fraction 16, which indicates a relative electrophoretic mobility of PSTV corresponding to a molecular weight of 5×10^4. The second major infectivity peak (fraction 20) corresponded to an apparent molecular weight of about 9.4×10^4 (Diener, 1971b).

Thus in both density gradient centrifugation and gel electrophoresis, more than one infectious species appeared to exist. The paucidisperse behavior of PSTV was considered to be due to the presence of aggregates of a minimal infectious unit (Diener, 1971b) and later work confirmed this conclusion (see below and Chapter 7).

In view of Loening's principle, the relatively close agreement between the molecular weight estimates of PSTV derived from sedimentation and gel electrophoretic analyses seemed to indicate that this estimate was not grossly in error. Later work, however, showed that this coincidence was fortuitous and that, because of the unique structure of PSTV, either method of analysis led to an underestimation of the molecular weight of PSTV (see Chapter 7).

The results nevertheless demonstrated unequivocally that the potato spindle tuber agent is an RNA of unexpectedly low molecular weight and that the agent, therefore, drastically differs from all known viral pathogens. Diener (1971b), on the basis of this profound difference, proposed the name *viroid* for PSTV and similar pathogens.

Further evidence for the low molecular weight of PSTV came from the observation that the RNA is able to enter gels of high polyacrylamide concentration (i.e., small pore size) from which high molecular weight

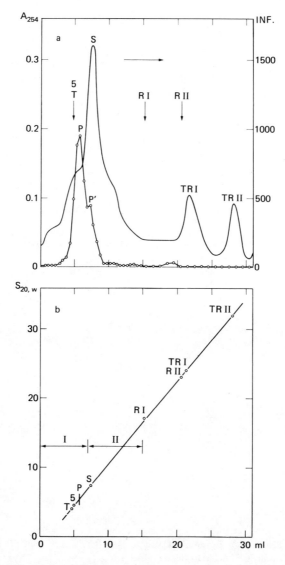

Figure 16. (a) Ultraviolet absorption profile (———) and infectivity distribution (0———0) of centrifuged sucrose density gradient containing PSTV (P and P'), S-TRSV-RNA (S), and the two TRSV-RNAs (TR I and II). Position in gradient of other marker RNAs is indicated by arrows: tRNA (T), rRNAs (R I and II), and 5 S RNA (5). Linear-log gradient, centrifugation for 26 hours at 24,000 rpm, SW 25.1 rotor, Spinco Model L centrifuge. Centrifugation is from left to right. A_{254} = absorbance at 254 nm; Inf. = infectivity index. (b) Relation between rate of sedimentation in gradient and $S_{20,w}$ values of marker RNAs. From: Diener (1971b).

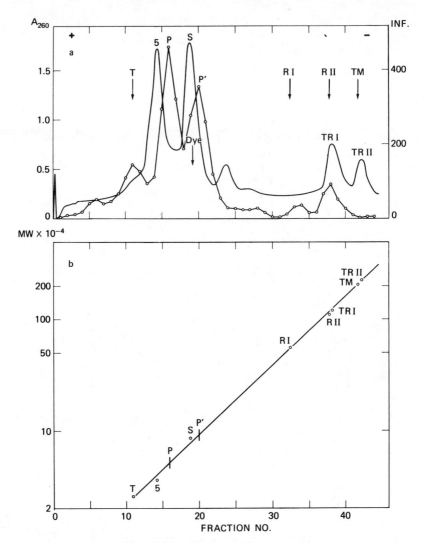

Figure 17. Electrophoretic mobilities of marker RNAs and of PSTV (P and P′) in a 3% polyacrylamide gel. (*a*) Ultraviolet absorption (————) and infectivity (0————0) profiles. A highly purified RNA preparation from PSTV-infected tissue, containing mainly 5 S RNA (5), to which S-TRSV RNA (S) and the two TRSV-RNAs (TR I and II) were added, was used. Position of other markers is indicated by arrows: tRNA (T), rRNAs (R I and II), and TMV-RNA (TM). A_{260} = absorbance at 260 nm; Inf. = infectivity index. (*b*) Relation between electrophoretic mobility and log molecular weight (MW) of marker RNAs. From: Diener (1971b).

RNAs are excluded. As shown in Figure 18, PSTV moves as a well-defined, single homogeneous band through such gels. Because *all* of PSTV is able to enter such gels (Diener and Smith, 1971), the results further strengthened the hypothesis that PSTV occurs in the form of aggregates of varying size which, upon application to gels of small pore size, disaggregate into the monomeric form of the infectious RNA.

4.1.5 Identification as Physical Entity

In all tests described so far, PSTV was identified by its biological activity, and no clearly recognizable ultraviolet light-absorbing component was correlated with infectivity distribution in density gradients

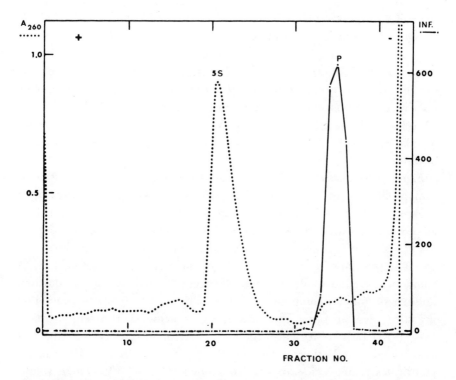

Figure 18.   Electrophoretic mobilities of 5 S ribosomal RNA (5S) and of PSTV (P) in a 20% polyacrylamide gel. Ultraviolet absorption (• • • •) and infectivity (• - • - •) profiles determined after electrophoresis for 4.5 hours at 4°C (5 mA per tube, constant current). A_{260} = Absorbance at 260 nm; Inf. = infectivity index. Electrophoretic movement from right to left. From: Diener and Smith (1971).

or polyacrylamide gels (Figs. 13, 14, 16, 17, and 18). Evidently, PSTV has an unusually high specific infectivity and occurs in infected plants in exceedingly small amounts. Further progress in the characterization of PSTV clearly required its isolation in purified form, followed by conventional biophysical and biochemical analyses. As a first step toward this goal, nucleic acids were extracted from relatively large amounts (5 kg) of healthy and PSTV-infected tomato leaves (Diener, 1972a). DNA was removed by treatment with DNase, high molecular weight RNAs by precipitation with lithium chloride, polysaccharides by extracting with methoxyethanol, and tRNA and DNA fragments by gel filtration in long columns of Sephadex G-100. PSTV was shown to elute with the excluded RNA which contained mainly 5 S ribosomal RNA (Diener. 1972a).

Analysis of these highly enriched low molecular weight RNA preparations in 20% polyacrylamide gels yielded the results shown in Figure 19.

Figure 19A shows the ultraviolet absorption profile of an RNA preparation from healthy tomato leaves after electrophoresis for 7.5 hours. Under the conditions used, 5 S RNA moved almost to the bottom of the gel. At least three additional ultraviolet-absorbing components with relative electrophoretic mobilities smaller than that of 5 S RNA are discernible in the gel (I, III, and IV). Identity of these RNA species is unknown; evidently they are minor low molecular weight components of cellular RNA.

Figure 19B shows the ultraviolet absorption and infectivity distribution profiles of an RNA preparation from PSTV-infected leaves after electrophoresis in a 20% polyacrylamide gel under identical conditions as those used with the preparation from healthy leaves. In addition to 5 S RNA, four ultraviolet-absorbing components are discernible in the gel. The positions in the gel of three of these (I, III, IV) coincide with components found in the preparations from healthy leaves. In addition, another prominent component is discernible (II). Bioassay of individual gel slices demonstrated that infectivity coincides with component II (Fig. 19B). This coincidence, the high level of infectivity, and the fact that component II does not occur in preparations from healthy leaves constitutes strong evidence that component II is PSTV. These results confirmed—by physical means—earlier conclusions that were based entirely on infectivity assays. Isolation of PSTV as an RNase-sensitive, ultraviolet-absorbing component with an absorption spectrum typical of

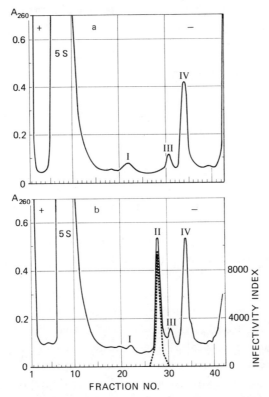

Figure 19. (a) Ultraviolet-absorption profile of RNA preparation from healthy tomato leaves after electrophoresis in a 20% polyacrylamide gel, for 7.5 hours at 4°C (5 mA per tube, constant current). (b) Ultraviolet-absorption (————) and infectivity distribution (•·····•) profiles of RNA preparation from PSTV-infected tomato leaves after electrophoresis in a 20% polyacrylamide gel [same conditions as in (a)]. 5 S = 5 S ribosomal RNA; I, III, IV = unidentified minor components of cellular RNA; II = PSTV; A_{260} = Absorbance at 260 nm. Electrophoretic movement from right to left. From: Diener (1972).

RNA demonstrated that the infectious agent is, indeed, an RNA. The observation that this RNA penetrates into 20% polyacrylamide gels and moves through such gels as a monodisperse, ultraviolet-absorbing band confirmed that the RNA is of very low molecular weight (Diener, 1972a). These observations laid the groundwork for the purification of PSTV and for the elucidation of a number of its properties (see Chapter 7).

4.2 CONFIRMING WORK WITH POTATO SPINDLE TUBER VIROID AND OTHER VIROIDS

4.2.1 Potato Spindle Tuber Viroid

Singh and Bagnall (1968) extracted infectious nucleic acid from PSTV-infected tissue and compared some of its properties with those of PSTV in crude extracts. They found that, compared with the agent in a buffer extract, the RNA had a greater infectivity dilution end point and higher thermal inactivation point, and was more sensitive to ribonuclease. Also, when infectivity of a buffer extract was removed by heating or by incubation with ribonuclease, it could be restored by treatment with phenol (Singh and Bagnall, 1968). From comparative centrifugation studies, the authors concluded that no free infectious RNA was present in buffer extracts, but did not specify what the RNA was bound to. They stated that in buffer extracts, the infectious nucleic acid "is influenced by some other factor" and suggested that "some of this influence may come from virus protein," but that "material (probably protein) in the sap of healthy plants could account for the inhibitory or protective effects that are the most obvious differences between the infectivity of [buffer extracts] and [nucleic acid preparations]" (Singh and Bagnall, 1968). The authors did not provide evidence for or against the presence of conventional viral nucleoprotein particles, but, on the contrary, they left the possibility open that PSTV was a more or less conventional virus.

Later, Singh and Clark (1971a, b) referred to the report of Diener (1970), in which the rate of sedimentation of PSTV in buffer extracts was given as about 10 S and stated that this was borne out by their "own preliminary studies using the Spinco Model E ultracentrifuge where we found a value of 5-6 S for the virus." The authors did not explain how this value was determined in the absence of purified PSTV, but they presented evidence obtained by density gradient centrifugation, followed by bioassay of individual fractions from the gradient, that PSTV sediments at very low rates (Singh and Clark, 1971b). The authors also presented the results of an analysis of the preparation in a 7.5% polyacrylamide gel which is purported to demonstrate that "peak infectivity corresponded to a particle size of 4-5 Svedberg units." The paucity of supporting evidence, however, makes it most unlikely that the claim of an "infectious low molecular weight RNA" could have been advanced solely on the basis of the results presented (Singh and Clark, 1971b).

4.2.2 Chrysanthemum Stunt Viroid

Lawson (1968a) reported that in buffer extracts of CSV-infected cineraria leaves, infectivity is associated with a fraction that only partially pellets during high-speed centrifugation and that sap extracts treated with butanol-chloroform and concentrated by ethanol precipitation were highly infectious. Phenol treatment of resuspended ethanol precipitates yielded samples with infectivities comparable to those of the untreated preparations. These properties of CSV resembled then known properties of PSTV in buffer extracts and suggested that the two pathogens might be of similar nature. Indeed, Hollings and Stone (1973) further studied the properties of CSV and suggested that CSV was "an uncoated RNA 'viroid.' " Definite proof of the viroid nature of CSV was presented by Diener and Lawson (1972, 1973), who demonstrated that partially purified CSV sediments in sucrose gradients with rates corresponding to S values of 5 to 7.5 (Fig. 20) and that CSV is able to enter 20% polyacrylamide gels, in which it moves with a mobility somewhat larger than that of PSTV (Fig. 21). Evidently, CSV, like PSTV, is a low mole-

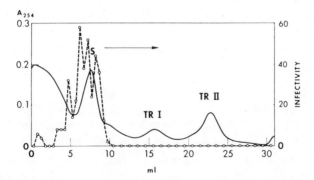

Figure 20. Ultraviolet-absorption profile (——) and infectivity distribution (0–0–0) of a sucrose density gradient containing partially purified CSV, tobacco ringspot satellite virus RNA (S), and the two RNAs of tobacco ringspot virus (TR I and TR II). Linear-log gradient, centrifugation for 26 hours at 24,000 rpm, SW 25.1 rotor, Spinco Model L Centrifuge. Centrifugation is from left to right. A_{254} = absorbance at 254 nm. Infectivity = average number of lesions on five half-leaves. From: Diener and Lawson (1973).

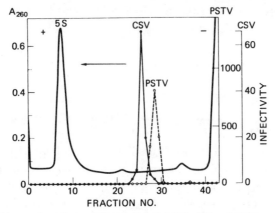

Figure 21. Electrophoretic mobilities of 5 S RNA (5S), of CSV, and of PSTV in a 20% polyacrylamide gel. Ultraviolet absorbance profile (————), infectivity distribution of CSV (—•—•—), and infectivity distribution of PSTV (0—0—0) determined after electrophoresis for 7.5 hours at 4°C (5 mA per tube, constant current). A_{260} = absorbance at 260 nm. Infectivity of PSTV = infectivity index, of CSV = average number of lesions on five half-leaves. Electrophoretic movement is from right to left. From: Diener and Lawson (1973).

cular weight RNA and, because its electrophoretic mobility is distinct from that of PSTV, CSV cannot be identical with PSTV.

4.2.3   Citrus Exocortis Viroid

Results similar to those of Diener and Raymer (1967) were reported by Semancik and Weathers (1968a, b), who investigated citrus exocortis disease. They found that most infectivity present in clarified extracts prepared from infected tissue remained in the supernatant solution after high-speed centrifugation; the authors came to the conclusion that the infectious entity possesses a sedimentation coefficient of about 10 to 15 S. Because of these results and because no typical virus particles could be observed in electron micrographs of infectious aliquots, the authors suggested that "the virus might represent an infectious nucleic acid form perhaps analogous to either the unstable form of tobacco rattle virus, the nonassembling mutants of tobacco mosaic virus, or a protein-free, double-stranded RNA molecule."

Later, Semancik and Weathers (1970a, b) reported that CEV is heterogeneous in size and density and that it exists in a major form sedimenting in the 10 to 15 S region and a minor form sedimenting at rates greater than 25 S. In contrast to the results of Diener and Raymer (1967, 1969) with PSTV, Semancik and Weathers (1970a, b) were unable to show partial resistance of CEV to RNase at high ionic strength of the incubation medium.[1] Also, the fact that the sedimentation rate of CEV could not be increased by melting and fast cooling led the authors to discount the possibility that CEV might be a double-stranded RNA (Semancik and Weathers, 1970a, b).

Later, however, Semancik and Weathers (1971) found that CEV was more slowly inactivated by diethylpyrocarbonate than was single-stranded viral RNA, that it eluted from methylated albumin at a buffer concentration (0.72 M) that is intermediate to that of DNA and double-stranded RNA, and that the infectivity distribution after cesium sulphate equilibrium sedimentation conformed to a broad density range (1.55 to 1.64 gcm^{-3}). The authors concluded that both single- and double-stranded RNA species were involved in CEV infectivity, but that the major infectious form was a "free-RNA double-stranded structure." However, based on essentially the same data, Semancik and Weathers (1972a) proposed two alternative hypotheses, namely that the infectious form of CEV might consist of a low molecular weight "tRNA-like" RNA or of an RNA:DNA hybrid molecule.

Later, Semancik and Weathers (1972b) stated that the buoyant density data of CEV obtained by equilibrium density gradient centrifugation were not reliable because low molecular weight nucleic acids had not reached equilibrium when samples were layered on top of the cesium sulphate column. When samples were mixed before centrifugation, some tRNA and CEV infectivity sedimented to a buoyant density of 1.63 gcm^{-3}, presumably suggesting that at least some of CEV consisted of double-stranded RNA.

In the same publication, the claim of a new species of "infectious" low molecular weight RNA as the causative agent of exocortis disease was advanced (Semancik and Weathers, 1972b). This claim was based on a reported coincidence in polyacrylamide gels between infectivity distribution and an A_{260} peak detectable only in nucleic acid prepara-

---

[1] Partial resistance of CEV to RNase at high ionic strength was, however, demonstrated by Semancik et al. (1975).

tions from CEV-infected plants. The results presented, however, were ambiguous because (1) preparations from infected plants sometimes contained two and sometimes three electrophoretic species migrating between 5 S RNA and the 14 S marker (Semancik and Weathers, 1972b; Figs. 2 to 5); (2) infectivity was in one case associated with the middle one of the three components (Fig. 2 of ref.), in another case located on both sides of the slowest migrating component, but not coincident with it (Fig. 3 of ref.), and in the third case illustrated only two A_{260} components were present and infectivity was located between the two (Fig. 4 of ref.). Furthermore, in a comparison of preparations from healthy and infected plants, the slowest-migrating A_{260} peak was absent in the extract from healthy plants. The authors recognized these discrepancies and commented that "the deviation from coincidence of the A_{260} peak and infectivity, especially in Figure 2, is probably due to the technical difficulties involved in scanning and cutting the same gel" (Semancik and Weathers, 1972b). It would appear, therefore, that the authors' claim of "the first association of 'pathogenic activity' with a physically discernible new RNA species" cannot be accepted in light of the evidence presented, particularly because in the same year coincidence of infectivity distribution in a gel with a new A_{260} peak had been demonstrated unequivocally with PSTV (see above).

Semancik and Weathers (1970a) tested the sensitivity of CEV in tissue extracts to exonucleases and reported that under conditions where tobacco rattle virus RNA was completely inactivated either by spleen or snake venom phosphodiesterase, CEV was still infectious. The resistance of CEV was, however, not complete, an observation that was explained by presumed endonuclease contamination of the enzyme preparations. Also, according to the authors, the lack of 3'- or 5'-hydroxyl groups at the termini could explain the resistance of CEV to exonucleases without the need for involving a circular structure. Later, the resistance of CEV to exonucleases was considered insufficient to postulate a circular structure for CEV (Semancik and Weathers, 1972a).

Sänger (1972), on the basis of its sedimentation, electrophoretic, and nuclease sensitivity properties, also concluded that CEV was a low molecular weight RNA; that is, a viroid.

The cumulative work with CEV, therefore, demonstrated that CEV had properties similar to those of PSTV, that a class relationship existed between the two agents, and that CEV had to be regarded as another example of a viroid.

4.2.4 Chrysanthemum Chlorotic Mottle Viroid

Romaine and Horst (1974, 1975) demonstrated that most of ChCMV could not be sedimented by high speed centrifugation, that SDS-phenol-treated preparations were generally more infectious than clarified buffer extracts, that ChCMV was sensitive to treatment with RNase, but not to treatment with DNase, and that the predominant infectious component sedimented in linear-log sucrose gradients with S values of between 6 and 14. These properties, together with the apparent absence of virus-like particles in infected tissue, strongly suggested that ChCMV is a viroid.

4.2.5 Cucumber Pale Fruit Viroid

Van Dorst and Peters (1974) referred to unpublished information demonstrating that CPFV is a viroid. Although this information has not so far been published, Sänger et al. (1976) provided such evidence by demonstrating that CPFV is a low molecular weight RNA with a structure and molecular weight similar to those of other viroids (see chapter 7).

4.2.6 Coconut Cadang-Cadang

Viroid etiology of cadang-cadang disease has not yet been established unequivocally, but two low molecular weight RNA species have been found in preparations from diseased palms but not in preparations from healthy ones (Randles, 1975). Furthermore, a comparative study with PSTV showed that the S_1 nuclease digestion kinetics of one of these RNAs, ccRNA-1, and PSTV are similar. The molecular weight of ccRNA-1 was estimated by gel electrophoretic analysis under denaturing conditions to be 63 to 73,000. In these and other properties, the structure of ccRNA-1 therefore appears to closely resemble that of viroids (Randles et al., 1976). Although transmission to young coconut seedlings, as judged by the appearance of characteristic symptoms of the disease. has now been achieved in two trials using either polyethylene glycol-precipitated material or total nucleic acid recovered from this fraction, inoculation with the viroidlike cc-RNA 1 alone has not, so far, led to disease symptoms (Randles et al., 1977). In both cases in which symptoms of cadang-cadang disease appeared, however, ccRNA-1 was detectable in nucleic acid extracts from the inoculated plants. Thus it appears likely that ccRNA-1 is involved in the etiology of cadang-cadang.

4.2.7 Hop Stunt Viroid

Sasaki and Shikata (1977b) characterized HSV as a viroid on the basis of its low sedimentation rate, sensitivity to treatment with RNase, insensitivity to treatment with DNase, phenol, or ethanol, and by the absence of viruslike particles in infected tissue.

5. DETERMINATION OF VIROID NATURE OF PATHOGEN OF UNKNOWN IDENTITY

To determine whether the agent of an infectious disease of unknown etiology has properties typical of viroids, several simple exploratory tests may be performed. By necessity, these tests are based on the properties of presently known viroids, and it must be stressed that newly discovered viroids may differ in some of their properties from the few viroids so far investigated. It is by no means certain, for example, whether all viroids are mechanically transmissible or whether all are composed of RNA. Also, depending on the host in which viroids replicate, they may exhibit different characteristics in crude extracts.

5.1 CRITERIA FOR SUSPECTING VIROID NATURE OF PATHOGEN

Viroid etiology of an infectious disease of unknown causation should be considered if all of the following observations have been made:

1. No microorganisms are consistently associated with the disease.
2. No viruslike particles are identifiable by electron microscopy of extracts or in sections from infected tissue.
3. Much of the pathogen in extracts cannot be pelleted by ultracentrifugation.
4. The agent is inactivated by either ribonuclease or deoxyribonuclease.

Evidently, the last two tests presuppose that the pathogen is mechanically transmissible and that a suitable assay host is available.

None of these observations constitute conclusive evidence for viroid etiology. The first two observations are negative evidence and improved techniques may lead to opposite conclusions. The third observation may indicate the presence of virus particles of low density, such as lipid-containing virions. The fourth observation could indicate the presence of virus particles with a loose protein shell that permits access of nucleases [such as cherry necrotic ringspot (Diener and Weaver, 1959), cucumber mosaic (Francki, 1968), or apple chlorotic leaf spot (Lister and Hadidi, 1971) viruses].

On the other hand, detection of viruslike particles in sections from infected tissue does not necessarily rule out viroid etiology of the disease in question; plants may be infected with a latent virus unrelated to the disease syndrome.

Also, some viroids are not readily released from host constituents or else they occur in extracts as aggregates of varying size (Diener, 1971a). In either case, much of the infectious material sediments faster than expected.

5.2 SEDIMENTATION PROPERTIES

Assuming that infectious extracts can be prepared from infected tissue and that a suitable bioassay host is available, accurate determination of the sedimentation properties of the infectious agent in such extracts should have high priority. This may conveniently be accomplished by subjecting an infectious extract, together with markers of known sedimentation coefficients, to velocity density-gradient centrifugation, followed by fractionation of the gradient and bioassay of all fractions. Conventional techniques of sucrose density-gradient centrifugation may be used, but precautions must be taken to assure freedom from nuclease contamination. Tissue extracts may be prepared in the presence of bentonite (ribonuclease inhibitor) and gradients should be made with ribonuclease-free sucrose. To obtain a reasonable estimate of infectivity distribution in the gradient, each fraction should be assayed undiluted and diluted 1/10 and 1/100.

Tissue extracts should be made with buffers of low and high ionic strength, such as 0.005 M and 0.5 M phosphate buffers, because some viroids are released from subcellular components only in high ionic strength medium (Raymer and Diener, 1969). Viroids usually are more stable in slightly alkaline than in acid media; thus buffers of pH 8 to 9 are recommended.

Extracts to be analyzed by density gradient centrifugation may be clarified by shaking with 1 volume of a 1:1 mixture (v/v) of chloroform and n-butanol, followed by centrifugation and withdrawal of the aqueous phase. This treatment does not adversely affect the infectivity of viroids, but results in much cleaner preparations.

A sedimentation coefficient of 7 to 15 S of the bulk of the infectious material suggests that the unknown pathogen may be a viroid. In crude preparations of this type, however, some infectivity often is found in lower portions of the gradient, and it is not uncommon to find low levels of infectivity in most fractions. For this reason, it is important to assay 10-fold dilutions of the inoculum and to determine where in the gradient the bulk of the infectious material is located.

5.3 NUCLEASE SENSITIVITY

All known viroids are composed of RNA and are inactivated by incubation with RNase. At high ionic strength, some viroids are partially RNase-resistant; thus incubation should be made in buffers of low and high ionic strength, such as 0.01 M and 0.5 M phosphate buffers, pH 7. Sensitivity should be investigated in the range of 0.1 to 1 $\mu g/ml$ of RNase with incubations for 1 hour at 25°C.

5.4 INSENSITIVITY TO TREATMENT WITH PHENOL

Treatment of buffer extracts from viroid-infected tissue with phenol or with phenol and SDS has little, if any, effect on the infectivity level of such preparations, or on the sedimentation properties of the infectious material. This is in contrast to extracts containing conventional viruses which after such treatments usually are much less infectious, and where the infectious material sediments at lower rates than before treatment with phenol.

To make these comparisons, 10 g of infected tissue are triturated in 20 ml of 0.5 M K_2HPO_4, 10 ml chloroform, and 10 ml n-butanol. The resulting emulsion is broken by low-speed centrifugation, and the aqueous phase is carefully withdrawn. One-half of the preparation is dialyzed versus 0.02 M phosphate buffer, pH 7, and, to the other half, 1 volume of water-saturated phenol is added. This preparation is shaken at room temperature for several minutes and is then centrifuged (20 minutes at 10,000 g). The aqueous (lower) phase is withdrawn and 2 volumes of ethanol are added. After storage for at least 30 minutes at 0°C, the sample is centrifuged (15 minutes at 10,000 g) and the supernatant is poured off. The pellet is resuspended in 0.02 M phosphate buffer, pH 7. After one or two more ethanol precipitations and resuspensions, the nucleic acid is finally suspended in the original volume of 0.02 M phosphate buffer, pH 7.

Both the phenol-treated and the nontreated samples are bioassayed at serial 10-fold dilutions, and the sedimentation properties of the infectious material are determined by density-gradient centrifugation in sister tubes and bioassay of all fractions.

5.5 ELECTROPHORETIC MOBILITY

If the above tests give results consistent with a viroid nature of the pathogen, infectious extracts should be subjected to polyacrylamide gel electrophoresis, as described in Chapter 6. This can be accomplished by analyzing partially purified preparations in gels of low and high concentration (such as 2.4% and 20% polyacrylamide gels), followed by cutting of the gels into 1- to 2-mm thick slices and bioassay of each slice after crushing in 0.02 M phosphate buffer and serial dilution.

In the 2.4% gel, peak infectivity should be in the region between tRNA and 16 S ribosomal RNA, indicating a low molecular weight of the infectious particles. The infectious particles should be able to enter 20% gels and to move through such gels as a sharp band.

Alternatively, preparations may be more thoroughly purified before electrophoresis and the gels stained to detect the suspected viroid (see Chapter 6).

6. PURIFICATION OF VIROIDS

Before the basic disparity between plant viruses and viroids had been recognized, several attempts were made to purify the putative virus responsible for the potato spindle tuber disease. In these efforts, methods customarily used for the purification of plant viruses were employed; in at least two cases isolation, or at least identification, of virus particles representing potato spindle tuber virus was claimed.

Allington *et al.* (1964) and Ball *et al.* (1964), on the basis of serological investigations, came to the conclusion that potato spindle tuber in Nebraska was caused by a strain of potato virus X (PVX) which, in contrast to other PVX strains, was able to multiply in the potato cultivar Saco. Bagnall (1967), however, could not detect any specific cross-reaction between antiserum believed to be against PSTV and PVX-containing sap, or between PVX antiserum and PSTV-containing sap. Furthermore, Raymer and Diener (1969) reported that in their source of PSTV, they were unable to find any evidence of PVX, either by plant indicator tests, electron microscopy, serology, or extraction and sedimentation studies. The results of Allington *et al.* (1964) could be explained by the assumption that PSTV and PVX were present as a mixture in their original source material and were not separated in subsequent transfers (Raymer and Diener, 1969).

Benson *et al.* (1964) and Singh *et al.* (1966) reported purification of potato spindle tuber virus from tomato by two procedures, one involving chloroform-charcoal clarification and the other repeated freezing and thawing of buffer extracts. With either method, further purification was achieved by differential low- and high-speed centrifugation, which yielded spherical particles with a predominant size of 25 nm. These preparations were infectious and the particles were considered to represent potato spindle tuber virus. These findings have not been confirmed and the identity of the spherical particles is in doubt (Raymer and Diener, 1969).

With the recognition of the unique properties of viroids, it became clear that their purification poses special problems: (1) No viral nucleo-protein particles are present. Thus no methods for the purification of conventional viruses are applicable. (2) Viroids occur in infected tissue in very small quantities. Their complete separation from normal cellular constituents, therefore, constitutes a formidable purification problem.

All methods of viroid purification developed so far consist in an application and, in some cases, modification of methods for the extraction, isolation, and separation of nucleic acids. In most cases, *all* nucleic acids are extracted from viroid-infected tissue and the viroid is then separated from one class of cellular nucleic acids after another until host nucleic acid can no longer be detected in the preparation.

In this chapter methods developed in the author's laboratory for the purification of PSTV are first described. Methods developed by other workers are then discussed and compared, and finally, methods for the assessment of purity of viroid preparations are considered.

6.1 BELTSVILLE METHODS

Two schemes for the purification of PSTV have been used in the author's laboratory with essentially identical yields of total low molecular weight nucleic acid. Both procedures involve (1) an initial homogenization in the presence of high ionic strength buffers (Diener and Raymer, 1969; Diener, 1971b, 1972a) and SDS to disrupt nucleoprotein complexes, (2) phenol extractions to remove protein, (3) extraction with ethylene glycol monomethyl ether to remove polysaccharides, (4) DNA digestion, (5) 2 M LiCl extraction to remove high molecular weight RNA, (6) gel filtration of total cellular low molecular weight RNA and DNA fragments, and (7) final purification by polyacrylamide gel electrophoresis. These two procedures are described in detail below for the purification of PSTV from 500 g of tomato leaf tissue. All operations are performed at 0°C to 4°C unless specifically stated otherwise; in all centrifugations, either Sorvall GSA rotors (large volumes) or SS-34 rotors (small volumes) were used.

6.1.1   Direct Phenol Extraction

6.1.1.1   Extraction of Nucleic Acids

Step a.   Combine 500 g frozen ($-20°C$) leaf tissue, 500 ml 1 M K_2HPO_4, 10 g SDS, 5 g bentonite, and 100 ml water-saturated phenol in 1-gallon stainless steel Waring Blendor container; grind 5 minutes at low speed.

Step b.   Centrifuge 10 minutes at 6500 rpm, remove aqueous (lower) phase by aspiration, and precipitate nucleic acids by adding 2 volumes cold 95% ethanol. Store overnight at $-30°C$.

Step c.   Carefully decant most of the upper ethanol phase before centrifuging 10 minutes at 6500 rpm. The nucleic acids are recovered as a tan "skin" at the interface between the lower phosphate-rich aqueous phase and the upper ethanol phase.

Step d.   Total nucleic acids are dissolved in 100 ml of water and reprecipitated with 2 volumes of cold 95% ethanol overnight at $-30°C$.

Step e.   Total nucleic acid is again recovered as a skin at the interface between the two phases after a 10-minute centrifugation at 6500 rpm. The nucleic acid is dissolved in approximately 40 ml of water.

6.1.1.2   Separation of Polysaccharides

Step f.   Polysaccharide extraction: (1) Add 1 volume of 2.5 M K_2HPO_4, 0.02 volume 85% H_3PO_4, and 1 volume ethylene glycol monoethyl ether. (2) Shake 2 minutes and centrifuge 5 minutes at 6500 rpm. (3) The upper phase, which contains the nucleic acid, is carefully removed without disturbing the interface and is dialyzed overnight against 2 liters of water, with at least one change during dialysis. (4) Nucleic acid is recovered by addition of 0.1 volume of 20% potassium acetate and 2 volumes of 95% ethanol and overnight storage at $-30°C$.

6.1.1.3   Separation of Major Cellular Nucleic Acids

Step g. DNA digestion: (1) Collect nucleic acid by 10-minute centrifugation at 6500 rpm. (2) Dissolve nucleic acid in approximately 30 ml water and add 3 μl M $MgCl_2$ and 0.01 ml 1 mg/ml DNase I (Worthington, electrophoretically purified) per 1 ml of nucleic acid solution. (3) Incubate 60 minutes at 25°C. (4) Add 0.025 volume 20% (w/v) SDS, 1 volume water-saturated phenol, and shake 5 minutes at room temperature. (5) Centrifuge 10 minutes at 6500 rpm, remove the upper aqueous phase, add 0.1 volume of 20% potassium acetate and 2 volumes of cold 95% ethanol, and precipitate overnight at − 30°C.

Step h. LiCl fractionation of RNA: (1) Collect RNA by 10-minute centrifugation at 6500 rpm and dissolve in approximately 30 ml of water. (2) Measure the volume of nucleic acid solution and add sufficient solid LiCl to yield a final concentration of 2 M (i.e., 84.8 mg/ml). Add the LiCl slowly while stirring the nucleic acid solution in an ice-water bath to prevent excessive heating. Let stand overnight at 4°C. (3) Centrifuge 10 minutes at 6500 rpm, save supernatant containing low molecular weight RNA. Add 2 M LiCl solution to the pellet, mix thoroughly, centrifuge as above, and save supernatant. Combine supernatants, add 2 volumes of cold 95% ethanol, mix thoroughly, and precipitate RNA overnight at − 30°C.

Step i. Collect the RNA by 10-minute centrifugation at 6500 rpm and dissolve it in 3 ml GM buffer (20 mM glycine, 3 mM $MgCl_2$, pH 9.0, with NaOH). Add 6 ml 95% ethanol and precipitate overnight at − 30°C.

Step j. Collect the RNA by 10-minute centrifugation at 6500 rpm and dissolve in 3 ml GM buffer for gel filtration.

6.1.2   Pretreatment with Chloroform-Butanol

The following method is a combination of earlier methods (Diener, 1971b) with a modification of the procedure of Morris and Wright (1975).

6.1.2.1   Extraction of Nucleic Acids

Step a.   Combine 500 g frozen ($-20°C$) tomato leaf tissue, 250 ml 16% (w/v) SDS, 250 ml GPS buffer (0.4 M glycine, 0.2 M Na_2HPO_4, 1.2 M NaCl, pH 9.5, with NaOH), 250 ml $CHCl_3$, and 250 ml *n*-butanol in a 1-gallon stainless steel Waring blendor container. Grind 3 minutes at high speed and recover aqueous (upper) phase by a 20-minute centrifugation at 8000 rpm.

Step b.   Extract the aqueous phase three times with phenol-$CHCl_3$. Add 500 ml water-saturated phenol and 500 ml $CHCl_3$ to the aqueous phase and stir 20 minutes at room temperature on a magnetic stirrer. Recover the aqueous phase by 10-minute centrifugation at 8000 rpm.

Step c.   Add 2 volumes of cold 95% ethanol to the aqueous phase and precipitate overnight at $-30°C$. Collect the nucleic acid precipitate by 20-minute centrifugation at 8000 rpm.

Step d.   Dissolve the precipitate in 150 ml TKM (10 mM Tris-HCl, pH 7.4; 10 mM $MgCl_2$) and dialyze overnight against 2 liters of TKM buffer, with at least one change of buffer during dialysis.

6.1.2.2   Separation of Major Cellular Nucleic Acids and Polysaccharides

Step e.   LiCl fractionation of nucleic acids: (1) Measure the volume of dialyzed nucleic acid solution and add sufficient solid LiCl to yield a final concentration of 2 M (i.e., 84.8 mg/ml). Add the LiCl slowly while stirring the nucleic acid solution in an ice-water bath to prevent excessive heating. Let stand overnight at 4°C. (2) Centrifuge 10 minutes at 6500 rpm and recover low molecular weight RNA and DNA from supernatant by adding 2.5 volumes of cold ethanol. Precipitate overnight at $-30°C$.

Step f.   Polysaccharide extraction: (1) Collect nucleic acid by 10-minute centrifugation at 8000 rpm and dissolve in 25 ml water. (2) Measure the volume and add 1 volume 2.5 M K_2HPO_4, 0.02 volume 85% H_3PO_4, and 1 volume ethylene glycol monomethyl ether. (3) Shake 2 minutes and centrifuge 5 minutes at 6500 rpm. (4) The upper phase, which contains the nucleic acid, is carefully removed and dialyzed overnight against 2 liters of water, with at least one change during dialysis. (5) The nucleic acid is recovered by addition of 0.1 volume of 2.2 M potassium acetate—0.1 M acetic acid, and 2.5 volumes of cold 95% ethanol. Precipitation overnight at $- 30°C$.

Step g. DNA digestion: (1) Collect nucleic acid by 10-minute centrifugation at 8000 rpm. Drain, dry the pellet with a N_2 stream, and dissolve in 10 ml water. Add 0.01 ml 1 M Tris-HCl, pH 7.5, 0.002 ml 1 M MgCl$_2$, and 0.01 ml 1 mg/ml DNase I per milliliter of nucleic acid solution. Incubate 60 minutes at 30°C. (2) Add 0.02 ml 500 mM Na$_2$EDTA, pH 8, 0.025 ml 20% (w/v) SDS, 0.1 ml Tris-HCl, pH 7.5, 0.5 ml water-saturated phenol, and 0.5 ml CHCl$_3$ per milliliter of solution. Shake 5 minutes at room temperature and centrifuge 10 minutes at 8000 rpm to recover the aqueous phase. (3) Precipitate nucleic acid by adding 0.1 volume 2.2 M potassium acetate—0.1 M acetic acid, and 2.5 volumes of cold 95% ethanol. Let stand overnight at $- 30°C$.

Step h. Collect the nucleic acid precipitate by 10-minute centrifugation at 8000 rpm. Dissolve in 1.5 ml GM buffer (20 mM glycine, 3 mM MgCl$_2$, pH 9.0, with NaOH) for gel filtration.

6.1.3 Gel Filtration

The next step in PSTV purification is gel filtration on long columns of either Sephadex or Ultrogel. We have found that gel filtration produces about an eight-fold enrichment of PSTV and a similar reduction in the numbers of polyacrylamide gels required for final purification.

Columns (1.6 × 170 cm) of either Sephadex G-100 (medium) or Ultrogel AcA44 (LKB Instruments) are prepared and equilibrated with GM buffer. Low molecular weight RNA prepared from 500 g of tomato leaf tissue (approximately 850 A_{260} units) is applied to the column in a volume of 1.5 to 3 ml. The column is eluted with GM buffer at a flow rate of 8 to 12 ml per hour. The effluent is continuously monitored at 280 nm (to keep the entire column profile on scale). Fractions of 2.5 to 4.0 ml are collected. A comparison of the fractionation obtained on Sephadex and Ultrogel is presented in Figure 22.

Polyacrylamide gel electrophoresis of samples across the column profiles shows that PSTV is located in the first peak of each column profile

—the peak marked 2 in the Sephadex profile and the peak marked 6 in the Ultrogel profile. Five S ribosomal RNA is found in peaks 3 and 7, respectively. Chromatography on Ultrogel clearly yields superior resolution of PSTV from 5 S ribosomal RNA.

Fractions containing PSTV are pooled and 0.1 volume of 2.2 M potassium acetate—0.1 M acetic acid, and 2.5 volumes of 95% ethanol are added. After overnight precipitation at $-30°C$, the RNA is collected by a 10-minute centrifugation at 8000 rpm, drained, dried with N_2, dissolved in 0.5 to 0.9 ml water, and stored at $-70°C$ before electrophoresis.

6.1.4   Gel Electrophoresis

Two electrophoresis procedures have proved satisfactory for the final purification of PSTV. Both involve extended electrophoresis on 20% acrylamide—0.5% Bis gels.

6.1.4.1   Cylindrical Gels

In the first procedure, the running buffer system of Loening (1967) is used (4.83 g Tris, 1.64 g sodium acetate, 0.37 g Na_2EDTA per liter, adjusted to pH 7.2 with glacial acetic acid). RNA from 500 g of tissue is dissolved in 0.45 ml H_2O, and 0.05 ml of 50% sucrose—5 times concentrated running buffer is added. Fifty-microliter aliquots are layered on top of each of ten 0.6 × 9 cm cylindrical gels and electrophoresed for 16 hours at 4 mA per tube. After electrophoresis, each gel is scanned at 260 nm in a Gilford spectrophotometer equipped with a linear transport. Gel segments containing PSTV are excised with a razor blade and stored frozen at $-70°C$ until a sufficient number of slices have been accumulated for RNA elution.

6.1.4.2   Slab Gels

The second electrophoresis procedure uses the running buffer system of Peacock and Dingman (1967) (10.8 g Tris, 5.5 g H_3BO_3, and 0.93 g Na_2EDTA per liter, pH 8.3). RNA from 500 g of tissue is dissolved in 0.9 ml H_2O, and 0.1 ml 50% sucrose—5 times concentrated running buffer is added. The sample is layered on a 10 (length) × 13 (width) × 0.24 (thickness) cm slab polymerized in the apparatus described by Studier (1973). After 60 minutes at 40 V, the voltage is increased to

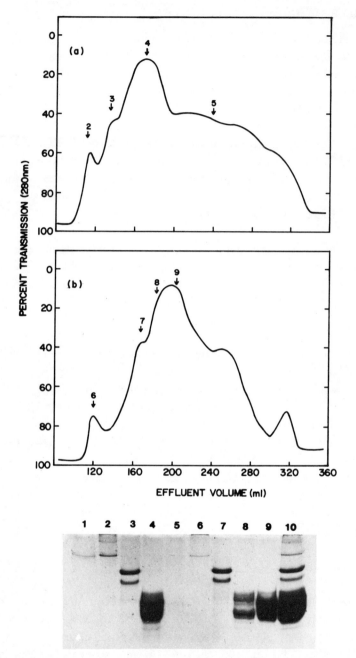

Figure 22. Fractionation of low molecular weight nucleic acids isolated from PSTV-infected tomato tissue by gel filtration. Total cellular RNA was prepared from 1 kg of frozen tomato seedlings tissue by the method using pretreatment with chloroform-butanol. Extraction with 2 M LiCl yielded 1710 A_{260} units of low molecular weight

130

100 V for the next 42 hours. The reservoir buffers are continuously recirculated. The PSTV zone is located by staining a thin strip cut from one side of the slab with 0.02% toluidine blue in 0.4 M sodium acetate —0.4 M acetic acid, excised, and stored at $-70°C$ until sufficient RNA has been accumulated for elution.

6.1.5 Recovery of Viroids from Gels

PSTV is eluted from polyacrylamide gels by a slight modification of the procedure of Diener (1973b). From 20 to 40 cylindrical gel slices or 4 to 6 strips from a slab gel are homogenized for 2 minutes at 0°C in 100 ml of GM buffer (20 mM glycine, 3 mM MgCl$_2$, pH 9.0, with NaOH) in a blender at full speed. The resulting slurry is centrifuged for 15 minutes at 500 g, and the sedimented gel particles are reextracted twice with 50 ml GM buffer. The combined supernatants are passed through a column of hydroxyapatite (Bio-gel HT, Bio-Rad Laboratories), previously equilibrated with GM buffer at room temperature. A column with 2-ml bed volume is sufficient to absorb 1 mg of nucleic acid. The flow rate should not exceed 2 ml per minute, and the effluent is monitored at 254 nm to detect any nucleic acid run-through.

After sample application, the column is washed with GM buffer until the effluent absorbance is essentially zero. PSTV is eluted as a sharp peak by washing the column with 20 mM glycine, 0.2 M K$_2$HPO$_4$ (pH 9.0, with NaOH). The PSTV solution is frozen, thawed, and centrifuged for 15 minutes at 10,000 g to remove insoluble material. Two volumes of cold 95% ethanol are added to the supernatant, and the RNA is allowed to precipitate overnight at $-20°C$. The PSTV is collected as a "skin" between the phosphate-rich lower phase and ethanol-rich upper phase following a 15-minute centrifugation at 5000 g. The drained PSTV is dissolved in a convenient volume of 1.25 M potassium phosphate,

---

nucleic acids. The nucleic acids were dissolved in 3 ml of GM buffer, and 1.5 ml aliquots were chromatographed on 1.6 × 170 cm (340 ml bed volume) columns of Sephadex G-100 (medium) and Ultrogel AcA44. Flow rates were 8 to 12 ml/hr. Two-hundred-microliter aliquots were removed from the indicated regions of the column profiles, 20 μl of 50% sucrose—100 mM Na$_2$EDTA, pH 8.1, were added, and the samples were electrophoresed on a 20% polyacrylamide slab gel in Tris-borate-EDTA running buffer for 16 hours at 4°C. Slot 1 contained 3.5 μg of purified PSTV; slot 10 contained 11.4 A$_{260}$ units of unfractionated low molecular weight nucleic acids. (a) Fractionation achieved with Sephadex G-100 (medium); (b) Fractionation achieved with Ultrogel AcA44. From: Diener et al. (1977).

pH 8.0, and extracted for 2 minutes with 0.5 volume of ethylene glycol monomethyl ether. The upper phase is carefully removed after a 5-minute centrifugation at 5000 g and is reextracted with the lower phase from a parallel mock extraction to remove remaining gel impurity. PSTV is recovered from the upper phase after dialysis against water by addition of 0.1 volume of 2.2 M potassium acetate—0.1 M acetic acid and 2.5 volumes of cold 95% ethanol and overnight storage at $-20°C$. PSTV is collected by a 15-minute centrifugation at 10,000 g, drained, dissolved in water, and stored frozen at $-70°C$. The ultraviolet spectrum of electrophoretically purified PSTV is typical of RNA with an absorbance maximum of 258 nm, a minimum of 230 nm, and a maximum: minimum ratio of 2.1 to 2.4. The extraction with ethylene glycol monomethyl ether may be repeated if the maximum:minimum ratio is low. The recovery of PSTV from acrylamide gels ranges from 46% to 91%.

6.1.6   Modification for Purification of Chrysanthemum Stunt Viroid

In contrast to extraction of tomato leaves which results in clear and colorless nucleic acid preparations, extraction of chrysanthemum leaves results in dark brown preparations, presumably because of high concentrations of phenolic compounds in chrysanthemum leaves. The following modification of the above purification procedure was effective in removing these compounds (Diener and Lawson, 1973).

After the second precipitation with ethanol, the pellet is resuspended in 35 ml of 1 M NaCl, 0.01 M Tris-HCl buffer, pH 7.8. Unlike PSTV preparations, CSV preparations at this stage of purification do not have ultraviolet spectra typical of nucleic acid. Furthermore, CSV preparations are deeply brown and viscous, whereas PSTV preparations are almost colorless.

To separate the impurities, 20 ml of commercial hydroxyapatite (suspended in 0.01 M sodium phosphate buffer) are added to the CSV preparation, and the mixtures are mechanically shaken at 4°C for 1 hour. The hydroxyapatite particles are then allowed to settle, and the supernatant solution is siphoned off. The hydroxyapatite particles are then washed twice with 10 ml each time of 1 M NaCl, 0.01 M Tris—HCl buffer, pH 7.8. Each time, the hydroxyapatite particles are allowed to settle and the supernatant solution is siphoned off. The supernatant solutions, which contain the colored impurities, are discarded, and the hydroxyapatite particles are poured into a chromatographic column (1 cm diameter). The column is washed with NaCl—Tris buffer until the effluent is devoid of absorbance at 254 nm. Nucleic acid is then eluted with 0.2 M K_2HPO_4—0.02 M glycine-HCl buffer, pH 9.0. The resulting preparations are colorless and have ultraviolet spectra typical of nucleic acids.

For further purification, the method described above is again followed, starting with step f, Section 6.1.1.2, or step e, Section 6.1.2.2. This procedure may be useful for the purification of viroids from other plant species containing large amounts of phenolic compounds.

6.2 OTHER PROCEDURES

Several other procedures for the purification of viroids have been published. These are described here insofar as they significantly vary from the procedures described above.

6.2.1 Citrus Exocortis Viroid Purification

Purification of CEV (Semancik and Weathers, 1972a; Semancik *et al.*, 1975) has several features in common with the procedure described. Fresh tissue from CEV-infected *Gynura aurantiaca* (10 g) is homogenized with 1 ml 5% SDS; 40 mg bentonite; 4 ml 0.1 M Tris-HCl, pH 8.9; 1 ml 0.1 M EDTA, pH 7.0; and 20 ml water-saturated phenol. The aqueous phase is recovered by centrifugation and is reextracted with water-saturated phenol. Nucleic acids are recovered by ethanol precipitation at 4°C. Polysaccharides are removed by ethylene glycol monomethyl ether extraction, and the ethylene glycol monomethyl ether phase is dialyzed against TKM buffer.

High molecular weight RNA is removed by addition of an equal volume of 4 M LiCl and subsequent centrifugation. The LiCl supernatant is gently mixed with 3 volumes of absolute ethanol and the windable nucleic acids are removed on a glass rod. The remaining precipitate is collected by centrifugation, digested with DNase, and reprecipitated with ethanol.

CEV in the low molecular weight RNA fraction adsorbs to CF-11 cellulose (Franklin, 1966) in the presence of either 35% or 15% ethanol in 0.1 M NaCl, 1 mM EDTA, 50 mM Tris-HCl, pH 7.2, and is eluted with the above buffer in the absence of ethanol (Semancik and Weathers, 1972a). The final steps in CEV purification are electrophoresis in 5% polyacrylamide gels, elution of the RNA from gel slices, and chromatography on columns of CF-11 cellulose-hydroxyapatite (1:1, w/w).

The major difference between this procedure for CEV purification and those previously described for PSTV purification is the use of CF-11 cellulose chromatography and electrophoresis in 5%, rather than 20%, polyacrylamide gels for final purification. Engelhardt (1972) has shown that chromatography on CF-11 cellulose can be used to discriminate between single-stranded RNA molecules in solution on the basis of their secondary structure. Elution in buffers that contain no ethanol is characteristic of double-stranded RNA and single-stranded RNA with a high degree of secondary structure. This behavior is consistent with the RNase resistance at high ionic strength and low-field nuclear magnetic resonance spectrum reported by Semancik et al. (1975). Although PSTV also shows a marked resistance to RNase digestion at high ionic strength (Diener and Raymer, 1969), its broad elution profile from CF-11 cellulose prevents the use of CF-11 cellulose chromatography in PSTV purification. Finally, electrophoresis in 20% rather than 5% polyacrylamide gels for final separation of PSTV results in better resolution of low molecular weight RNA (Diener et al., 1977).

Morris and Wright (1975) modified and simplified an earlier procedure for the purification of CEV (Semancik and Weathers, 1968b) and used it as a diagnostic method for the detection of PSTV in potato "seed." Their homogenization buffer contains 0.2 M Na_2SO_3 and 0.1% sodium diethyldithiocarbamate to minimize the formation of dark brown pig-- ments by oxidation of plant phenolic compounds. These additions, however, have no noticeable effect when PSTV is isolated from tomato seedlings (Diener et al., 1977). Because the procedure of Morris and Wright is intended to be used to diagnose potato spindle tuber disease, only small amounts of tissue are processed. For this reason, the polysaccharide extraction, DNA digestion, and gel filtration steps can be omitted. No doubt this method would be suitable also for the large-scale purification of viroids, provided the omitted steps are reintroduced.

Another method for the extraction of CEV has been published by Sänger (1972). This method consists of a standard extraction of nucleic acids with pH 7.5 buffer (0.1 M Tris-HCl, 0.01 M EDTA, 0.01 M mercaptoethanol, 0.1 M NaCl)—phenol and SDS in the presence of bentonite. The aqueous phase is retreated with phenol and bentonite, and nucleic acids are precipitated once with cetyltrimethylammoniumbromide and then three more times with ethanol.

Later this method was extended by addition of lithium chloride fractionation and polyacrylamide gel electrophoresis as a final purification step (Sänger and Ramm, 1975). Still later, the method was further modified to include a step to remove polysaccharides (in essence identical with step f, Section 6.1.1.2) (Singh and Sänger, 1976). The final method was used for the purification of PSTV as well as of CEV.

Further purification of several viroids has been reported by Sänger et al. (1976) with a modification of the above extraction method, followed by preparative gel electrophoresis. The stained viroid band was cut out and extracted from the gel by homogenization in 0.1 M Tris-HCl (pH 7.5), 0.1 M NaCl, and 0.1 mM EDTA in the presence of 20% phenol, adsorbed on DEAE-cellulose, desorbed with 1.2 M NaCl, and precipitated with ethanol. The electrophoretic separation and purification was repeated three times.

6.2.2 Cadang-Cadang Ribonucleic Acid

Randles (1975) extracted the cadang-cadang disease-associated low molecular weight RNAs (see Chapter 4) by chopping leaflet tissue from affected palms in 0.1 M Na$_2$HPO$_4$ containing 0.1% sodium thioglycollate and 0.01 M sodium diethyldithiocarbamate, straining through muslin, and centrifuging at low speed. To precipitate macromolecules, solid polyethylene glycol (PEG 6000) is added to the supernatant (to a final concentration of 5%), the mixture is stirred in the cold, and after 1 to 2 hours, the precipitate is collected by low-speed centrifugation. The precipitates are resuspended in 0.05 M sodium phosphate buffer, pH 7.2, and nucleic acids are extracted from this by conventional phenol-SDS treatment, followed by ethanol precipitation of the nucleic acid. After incubation of the preparations with Streptomyces griseus protease and another phenol-SDS treatment and ethanol precipitation, the preparations are subjected to polyacrylamide gel electrophoresis, and the disease-associated RNAs are recovered from the gels by phenol-SDS extraction (Randles, 1975).

Attempts to adapt this procedure to PSTV revealed that, in contrast to the cadang-cadang-associated RNAs, PSTV is not completely precipitated by addition of polyethylene glycol (Diener et al., 1977).

6.3 ASSESSMENT OF PURITY

Before the purified viroid is used in any physical or chemical analysis, it is of obvious importance to ascertain the degree of purification achieved.

Most conveniently, this is achieved by subjecting the purified viroid preparation to another cycle of gel electrophoresis, followed by determination of the infectivity distribution in the gel.

In the original purification of PSTV (Diener, 1972a; see Section 6.1), analysis of the final product in a 20% polyacrylamide gel and bioassay of individual gel slices revealed excellent coincidence between the major ultraviolet absorbing component discernible in the gel and infectivity distribution (Fig. 23). Some contamination with what appears to be 5 S RNA, however, was evident (Fractions 9 to 12). No such contamination was discernible after two cycles of gel electrophoresis and in later PSTV preparations (Diener *et al.*, 1977).

Observation in gels of a single component does not necessarily mean purity of the viroid preparation, because a normal RNA species may comigrate with the viroid. This often occurs, for example, when low molecular weight RNA preparations from PSTV-infected plants are analyzed in 10% polyacrylamide gels, in which "9 S" RNA (which is a

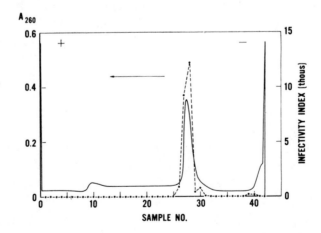

Figure 23.   Ultraviolet-absorption (———) and infectivity distribution (• - - - •) profiles of purified PSTV after electrophoresis in a 20% polyacrylamide gel for 7.5 hours at 4°C (5 mA per tube, constant current); A_{260} = absorbance at 260 nm. Electrophoretic movement from right to left. From: Diener and Smith (1973).

normal component of cellular RNA) comigrates or nearly comigrates with PSTV (Diener *et al.*, 1977).

To investigate this possibility, two approaches are useful. (1) The preparation should be analyzed in gels of different acrylamide concentration, because in these the normal constituent and the viroid may separate during electrophoresis. This is the case, for example, with "9 S" RNA and PSTV. The two RNAs are readily separated by electrophoresis in 20% polyacrylamide gels (Diener, 1972a). (2) Preparation of a low molecular weight RNA preparation from healthy plants in a manner identical with that used for viroid purification and analysis, in gels, together with the viroid preparation may disclose a cellular RNA species that comigrates with the viroid in the particular system used. For example, normal RNA that comigrates with cadang-cadang-associated RNA was discovered in a comparison of low molecular weight RNAs from normal and cadang-cadang-affected palms (Randles *et. al.*, 1976). In this case, it could be shown that the thermal denaturation properties of the normal RNA are typical of those of single-stranded RNAs, whereas thermal denaturation of the cadang-cadang-associated RNA is typical of that of viroids (Randles *et al.*, 1976).

Furthermore, two-dimensional RNA fingerprints of pancreatic RNase and ribonuclease T_1 digests of ^{125}I-labeled normal and cadang-cadang-associated RNAs conclusively showed that the latter RNA is not detectable in preparations from healthy tissue and that primary sequences of the two RNAs are distinct (Dickson, 1976). RNA fingerprint analysis, then, is a sensitive method for the assessment of viroid purity.

7. PHYSICAL PROPERTIES OF VIROIDS

With the development of methods for the purification of viroids, it became possible to determine some of their physical and chemical properties. Here, we are concerned with physical characteristics of viroids that have been investigated to date. These properties mainly concern the size (molecular weight) and shape (molecular structure) of viroids.

7.1 SIZE (MOLECULAR WEIGHT)

As described in Chapter 4, the low molecular weight of the first recognized viroid, PSTV, had already been demonstrated conclusively before it was available in pure form, and even before it was recognizable as a physical entity.

7.1.1 Determination by Gel Electrophoresis

With the availability of purified PSTV, a redetermination of its molecular weight, based not on its biological activity but on its absorption of ultraviolet light, became possible. For this purpose a method described by Boedtker (1971) appeared particularly promising, because this method was supposed to permit the determination of the molecular weights of RNAs independently of their conformations.

RNAs are treated with formaldehyde at 63°C, a treatment which, according to Boedtker (1971), completely denatures the RNAs; that is, destroys their secondary and tertiary structures. Consequently, it is believed that the electrophoretic mobility of formylated RNAs in polyacrylamide gels is a function of their molecular weights, but not of their particular native conformations. Because PSTV is inactivated by exposure to formaldehyde, this method could not be used as long as the only available parameter for detection of PSTV was its biological activity.

Determination of the electrophoretic migration rate of formylated PSTV as compared to that of similarly formylated RNA markers of known molecular weight, resulted in a molecular weight estimate for PSTV of 7.5 to 8.5 $\times$ 10^4 (Diener and Smith, 1973). The pronounced discrepancy between this value and that obtained with native PSTV (5×10^4) (see Chapter 4) is most likely due to the compact molecular structure of native PSTV (see Section 7.5) and illustrates the substantial error inherent in efforts to estimate molecular weights of nucleic acids of unknown conformation by gel electrophoresis of native molecules.

Semancik et al. (1973b) estimated the molecular weight of CEV and PSTV in 5% polyacrylamide gels under nondenaturing conditions. They obtained values of 5×10^4 or 1×10^5 daltons with either viroid, depending on acrylamide concentration, but considered the former value in error because of aberrant migration of the viroids in gels of high acrylamide concentration. The investigators preferred a molecular weight estimate of 1.25×10^5 for either viroid (Semancik et al., 1973b). Sänger (1972) similarly studied the electrophoretic mobility of native CEV. He estimated its molecular weight as 5 to 6 $\times$ 10^4, however. It is now clear that unambiguous molecular weight determinations by gel electrophoresis were frustrated by the unknown and evidently unique structure of viroids that rendered all RNA standards of known molecular weight inappropriate.

7.1.2 Determination by Electron Microscopy

Direct length measurement of native PSTV molecules in electron micrographs (see Section 7.2.1.4) yielded a molecular weight estimate of 8.9×10^4 for PSTV (Sogo et al., 1973). Evidently, this value closely agreed with the values obtained by gel electrophoresis of formylated PSTV (Diener and Smith, 1973).

More recently, molecular weight estimates of PSTV have been obtained by length measurements of fully denatured molecules visualized by electron microscopy. According to McClements and Kaesberg (1977), the length of linear, denatured PSTV is 110 ± 10 nm and that of circular denatured PSTV (see Section 7.2.1.4) is 140 ± 10 nm. On the assumption that the linear density of PSTV is the same as that determined for brome mosaic virus RNAs (1.20×10^6 daltons/μm) the lengths of PSTV molecules translate into molecular weights of 1.1×10^5 for the linear and 1.37×10^5 for the circular PSTV molecules (McClements and Kaesberg, 1977).

7.1.3 Determination by Irradiation

Theoretically, the smaller a nucleic acid (i.e., the smaller the target size), the more resistant it is to inactivation by ultraviolet or ionizing radiation. Thus it should be possible to estimate the size of an infectious nucleic acid by determining the dose of radiation necessary to inactivate it. Although this is possible with ionizing radiation, the effect of size on ultraviolet sensitivity of nucleic acids is not as well understood (Adams, 1970). So far, the effect of ultraviolet radiation on PSTV and that of ionizing radiation on CEV have been investigated. Figure 24 shows that exposure of purified PSTV, of tobacco ringspot virus, or of its satellite (Schneider, 1971) to ultraviolet radiation of 254 nm results in an inactivation dose for PSTV and satellite virus that is 70 to 90 times as large as that for tobacco ringspot virus (Diener et al., 1974). In view of the fact that both PSTV and the RNA contained in satellite virus are low molecular weight RNA species, it appears likely that this marked difference is a consequence of the much smaller size (smaller target volume) of PSTV and satellite RNA, as compared with tobacco ringspot virus RNA.

Semancik et al. (1973b) exposed preparations of CEV and tobacco mosaic virus to ionizing radiation and determined from comparative rates of biological inactivation a target size of 1.1×10^5 daltons for CEV.

Both the high resistance of PSTV to ultraviolet radiation and the small target size of CEV confirm the low molecular weights of the respective viroids.

7.1.4 Determination by Equilibrium Sedimentation

Far more dependable molecular weight estimates have now been obtained by high-speed (15,000 rpm) as well as low-speed (7000 to 8000 rpm)

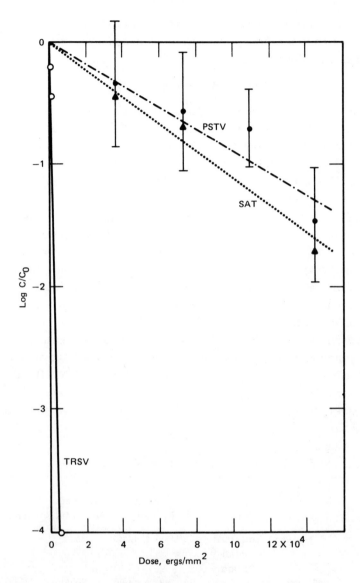

Figure 24. Inactivation of PSTV, SAT, and TRSV by ultraviolet light of 254 nm. Logarithm of survival ratios (C/C_0) plotted as a function of incident dose. •, Purified PSTV; vertical bars, 95% fiducial limits for ratios to controls; ▲, purified SAT; o, purified TRSV. From: Diener *et al.* (1974).

equilibrium sedimentation in an analytical ultracentrifuge (Sänger *et al.*, 1976). The concentration profiles were evaluated from the 'formula for monodisperse solutions:

$$\Delta \ln c = [M(1 - \bar{v}{\bullet}\rho)\omega^2/2RT]{\bullet}\Delta r^2$$

where M is the molecular weight, $\bar{v}$ is the partial specific volume of the solute, ρ is the density of the solution, R is the molar gas constant, T is the absolute temperature, ω ist the angular speed of the rotor, c is the concentration of the solute, and r is the radial distance from the axis of rotation. The concentration profiles, when plotted according to $\Delta \ln c$ against Δr^2 and evaluated over a concentration range of more than an order of magnitude, gave straight lines with high accuracy and led to the molecular weight estimates shown in Table 24 (Sänger *et al.*, 1976).

These values are based on a partial specific volume of $\bar{v} = 0.53$ cm³/g which is the known value for the sodium salt of a crude mixture of tRNA from yeast and which, according to the authors, can be safely used for viroids because (1) the degree of base-pairing and the composition of nucleotides are comparable in CEV and tRNA, (2) the particular tertiary structure of tRNA has only minor influence on $\bar{v}$, and (3) very similar values of $\bar{v}$ have been determined for the sodium salt of 16 S ribosomal RNA from *Escherichia coli* and for RNA from brome mosaic virus (Sänger *et al.*, 1976).

According to these results, molecular weights of the examined viroids

Table 24.   Molecular Weight Estimates of Viroids

Viroid	Source Plant	Method	Molecular Weight Estimate	Calculated Number of Nucleotides	References[a]
CEV	*Gynura*	Equilibrium sedimentation	119,000 ± 4,000	357[b]	1
CEV	Tomato	"	119,000 ± 4,000	357[b]	1
CPFV	Tomato	"	110,000 ± 5,000	330[b]	1
PSTV	Tomato	"	127,000 ± 4,000	381[b]	1
PSTV	Tomato	Sequence	115,500[c]	359	2

[a] References: 1 = Sänger *et al.* (1976); 2 = Gross *et al.* (1978).
[b] Based on average molecular weight of 322 per nucleotide plus an average value of 11 per nucleotide from bound Mg^{2+} and Na^+ ions (Sänger *et al.*, 1976).
[c] Domdey *et al.* (1978).

do not greatly differ one from the other, but the differences found correspond to significant differences in chain length between the smallest (CPFV) and the largest (PSTV) viroid investigated.

7.1.5   Determination from Nucleotide Sequence

On the basis of the nucleotide sequence of PSTV (Domdey *et al.*, 1978; Gross *et al.*, 1978), resulting in an estimate of about 359 nucleotides in the viroid molecule, the molecular weight of PSTV has been estimated as about 115,500 (Domdey *et al.*, 1978) (Table 24).

7.2   SHAPE (MOLECULAR STRUCTURE)

Certain clues as to the molecular structure of PSTV were obtained already before viroids could be recognized as physical entities. For a number of years, however, the molecular structure of viroids remained enigmatic because, from results obtained with different analytical systems, different and often contradictory conclusions seemed to be called for. These uncertainties revolved primarily around two questions: whether viroids are single- or double-stranded RNA molecules and whether these molecules are linear or circular structures.

7.2.1   Single- or Double-Stranded Ribonucleic Acid

As has been noted in Chapter 4, PSTV is partially resistant to inactivation by RNase if incubated in a medium of relatively high ionic strength. These observations led to the hypothesis that PSTV may be a double-stranded RNA molecule (Diener and Raymer, 1967). When PSTV was later analyzed in various analytical systems, some results appeared to confirm, others to invalidate, this hypothesis.

7.2.1.1   Chromatographic Properties

The elution pattern of PSTV from columns of methylated serum albumin suggested double-strandedness of the RNA (Fig. 25) because PSTV eluted primarily in those fractions that contained the host DNA or immediately followed DNA (Diener and Raymer, 1967, 1969).

From columns of CF-11 cellulose (Franklin, 1966), however, PSTV

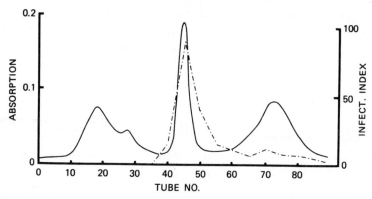

Figure 25. Elution patterns from a MAK column of a phenol-treated PSTV concentrate. (——————), absorbance profile; (- - - - -), infectivity distribution. One milligram of nucleic acid (in 0.6 ml of 0.4 M saline-phosphate buffer) was applied to the column. Flow rate, 0.8 ml/min, 0.4 to 1.2 M saline-phosphate. From: Diener and Raymer (1969).

eluted both in 15% ethanolic and in ethanol-free buffer (Fig. 26), a result that is compatible with the presence of both single- and double-stranded infectious molecules (Diener and Raymer, 1969). A possible explanation for these observations may be deduced, however, from the work of Engelhardt (1972), who showed that the extent of secondary structure of an RNA has a profound influence on its elution pattern from such columns. The greater the amount of secondary structure of an RNA at the time of addition to the column, the greater is the fraction that will elute in ethanol-free buffer; that is, in the eluate that was formerly believed to consist solely of double-stranded RNA (Franklin, 1966). Judged by this criterion, native PSTV has an extensive secondary structure, but it may not be a perfectly base-paired duplex molecule.

From hydroxyapatite, on the other hand, PSTV elutes mostly at a phosphate buffer concentration lower than expected of a double-stranded RNA (Fig. 27) (Diener, 1971c; Lewandowski *et al.*, 1971). Some PSTV, however, elutes at higher buffer concentration (Diener, 1971c).

CEV, like PSTV, elutes from columns of methylated serum albumin in the range of host DNA, but, unlike PSTV, CEV elutes from cellulose columns almost exclusively in ethanol-free buffer (Semancik and Weathers, 1970a). Although these properties of CEV are compatible with a double-stranded structure, attempts to melt the suspected duplex RNA or to observe any resistance to RNase in a high ionic strength

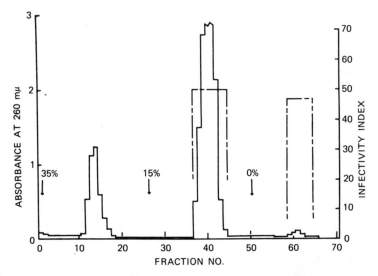

Figure 26.   Absorbance profile and infectivity distribution in eluate from a cellulose column of a phenol- and DNase-treated PSTV concentrate. Two milligrams of nucleic acid (in 0.92 ml of STE-35% ethanol) was applied to the column. Arrows indicate change of eluting buffer, 35% = STE-35% ethanol buffer; 15% = STE-15% ethanol buffer; 0% = STE buffer. From: Diener and Raymer (1969).

medium were unsuccessful (see Chapter 4). Later, however, the authors did achieve a suggestion of strand separation by heating and rapid cooling (Semancik and Weathers, 1972a). They furthermore determined that CEV is partially resistant to incubation with diethylpyrocarbonate, a result that parallels those of Singh and Clark (1971b) with PSTV and suggests a structure that is at least partially double-stranded (Öberg, 1970). Contrary to expectations, however, CEV is readily inactivated by formaldehyde (Semancik and Weathers, 1972a).

It is evident that determination of chromatographic properties alone is not sufficient to elucidate the molecular structure of viroids.

7.2.1.2   Immunological Properties

Stollar and Diener (1971) made immunological tests with antisera that react specifically with double-stranded RNA (Schwartz and Stollar, 1969). As shown in Figure 28, no evidence was found for the presence of double-stranded RNA in highly infectious PSTV preparations, thus indicating that PSTV was not recognized by these antisera as a double-stranded structure.

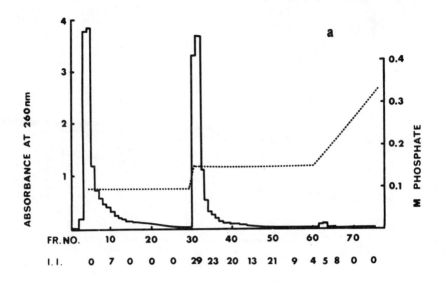

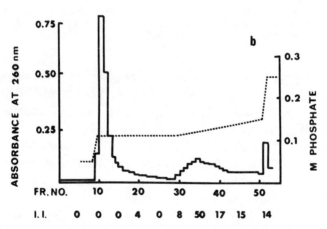

Figure 27. Absorbance profiles and infectivity distributions of eluates from hydroxy-apatite columns. (*a*) Elution profile of a phenol- and DNase-treated PSTV preparation. (*b*) Elution profile of the pooled, reconcentrated, and dialyzed 0.15 *M* phosphate buffer eluate of (*a*). (————), absorbance profiles; (- - - - -), molarity of phosphate buffer; Fr. No. = fraction number; I.I. = infectivity indexes of individual fractions. From: Diener (1971c).

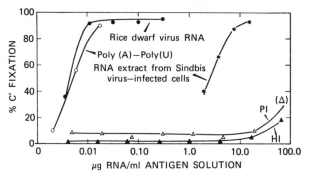

Figure 28. Complement (C′) fixation of antipoly (A) poly (U) rabbit serum with varying quantities of rice dwarf virus RNA; poly (A) poly (U); RNA extract from Sindbis virus-infected cell; RNA extract from healthy tomato plants (HI); and RNA extract from potato spindle tuber viroid-infected tomato plants (PI). (△), value corrected for anticomplementariness of antigen. From: Stollar and Diener (1971).

7.2.1.3 Thermal Denaturation Properties

More definitive results regarding the molecular structure of viroids were achieved once purified PSTV became available and the thermal denaturation properties of the RNA could be determined. In $0.01 \times SSC$ (SSC = $0.15\ M$ NaCl—$0.015\ M$ sodium citrate, pH 7.0), the total hyperchromic shift of PSTV was found to be about 24% and the t_m (temperature at which 50% of the molecules are denatured) was about 50°C (Diener, 1972a). The thermal denaturation curve (Fig. 29) indicates that PSTV is not a regularly base-paired structure, such as double-stranded RNA, because in this case denaturation would be expected to occur at higher temperature, as indicated in Figure 29 by the thermal denaturation curve of a genuine double-stranded RNA, rice dwarf virus RNA. The relatively narrow temperature range for denaturation, however, suggests the presence of extensive regions of intramolecular complementarity.

At higher ionic strength ($0.1 \times SSC$) the thermal denaturation curve of PSTV was similar, except that under these conditions, t_m was about 54°C (Diener and Hadidi, 1977).

Semancik *et al.* (1975) reported that CEV in $0.1 \times SSC$ has a t_m of 52°C with a hyperchromic shift of about 22%, thus confirming the earlier results obtained with purified PSTV (Diener, 1972a). The authors commented that the hyperchromicity profile of CEV is intermediate

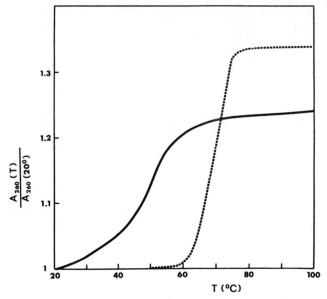

Figure 29.   Thermal denaturation curves of PSTV (———) and of double-stranded rice dwarf virus RNA (- - - - -) in 0.01 × SSC.

between tRNA[Phe] and double-stranded RNA. The reannealing profile of CEV indicated that the RNA rapidly regains its molecular structure upon cooling. As was found for PSTV, denaturation of CEV occurs within a relatively narrow range of temperatures, indicating that the extent of base-pairing in the molecule is larger than that in tRNA (about 50%), but less than in genuine double-stranded RNA. The authors also demonstrated that the addition of low concentrations of Mg^{2+} or 1 × SSC salt greatly stabilizes CEV to thermal denaturation. Thus with 5 mM Mg^{2+} added to 0.1 × SSC or in 1 × SSC, the t_m of CEV was raised to 79°C (Semancik *et al.*, 1975).

Sänger *et al.* (1976) reported that the t_m values of three viroids examined (CEV, PSTV, and CPFV) in 0.01 M sodium cacodylate, pH 6.8, 1 mM EDTA were identical (about 51°C) and the hyperchromic shift was 22%. Again, denaturation occurred over a narrow temperature range, indicating a high degree of base-pairing (Sänger *et al.*, 1976).

Rather different results were obtained by Randles (1975) for the smallest cadang-cadang-associated RNA (ccRNA 1). The hyperchromic shift amounted to only about 10% and the t_m was about 58°C when determined in 0.1 × SSC.

On the basis of quantitative thermodynamic and kinetic studies of the thermal denaturation of CPFV, Henco *et al.* (1977) developed a tentative model of the molecular structure of this viroid. This model involved the existence of an uninterrupted double helix of 52 base pairs, flanked on either side by mismatched regions and single-stranded loops separated by short base-paired regions.

The presence of a long, uninterrupted double helix in PSTV or CEV would be inconsistent with the results of experiments in which treatment of unlabeled PSTV or ^{125}I-labeled CEV preparations with double-strand-specific *E. coli* RNase III affected neither the electrophoretic mobility of PSTV or ^{125}I-CEV (Dickson, 1976) nor the infectivity of PSTV (Diener, unpublished). For RNA to be cleaved by RNase III, it must contain either an extended region of perfect double-stranded RNA (25 or more base pairs) or a highly specialized RNA sequence (Robertson and Hunter, 1975); the observed lack of effect implies that neither viroid contains such regions.

In a further study of the thermal denaturation properties of viroids (Langowski *et al.*, 1978), a refinement of the earlier model was proposed. In this model, viroids exist in their native conformation as extended rod-like structures characterized by a series of double helical sections and internal loops. In the different viroid species 250 to 300 nucleotides out of a total 350 nucleotides are needed to interpret the thermodynamic properties of the molecules (Langowski *et al.*, 1978). In contrast to the earlier model, which proposed a long, uninterrupted double-helical region, the refined model proposes that each helical sequence of four to five base pairs is followed by a defect on the average in form of an internal loop of two bases (Langowski *et al.*, 1978). Evidently, this model is consistent with the known resistance of viroids to RNase III.

In a calorimetric study of viroids, Klump *et al.* (1978) deduced from the total enthalpy that 85 base pairs exist in their native structure in a series of helical sections and internal loops.

7.2.1.4 Visualization of Native Viroids

Sogo *et al.* (1973) were the first investigators to achieve visualization of a native viroid by electron miscroscopy.

Purified PSTV preparations were processed for electron microscopy by the protein monolayer spreading technique of Kleinschmidt and Zahn (1959). When PSTV preparations in 4 *M* sodium acetate were spread onto a hypophase of distilled water, very short structures, mostly in compact aggregates but occasionally as separate particles, were revealed. These structures could not be detected in nucleic acid-free controls, and they were absent from preparations treated with ribonuclease.

Because of the association of PSTV molecules in aggregates, conclusions as to their length and structure were difficult to draw from micrographs obtained with this method. Consequently, methods that promised to dissociate the aggregates and to make possible the visualization of individual PSTV molecules were investigated.

The method of Granboulan and Scherrer (1969) appeared suitable, since 8 M urea is known to suppress weak bonding forces. As shown in Figure 30a, disaggregation indeed occurred with this method, and large numbers of individual short strands of relatively uniform length were visible. No such strands were discernible in the control samples or after treatment with ribonuclease (Fig. 30c).

To determine whether heating of PSTV in 8 M urea leads to unfolding of the molecules, preparations that had been heated for 10 minutes at 62°C, followed by quenching in ice water, were examined (Fig. 30b). Length measurements of over 2400 PSTV molecules showed, however, that the lengths of unheated and heated molecules did not differ significantly. Average length of molecules was found to be about 50 nm (Fig. 31).

Since the mass per unit length is unknown, the lengths of PSTV molecules were compared with those of nucleic acids of known molecular weight, which were added to PSTV preparations and which were thus treated identically.

The frontispiece shows an electron micrograph of a mixture of a double stranded DNA, namely coliphage T_7-DNA, and PSTV. Assuming a molecular weight of T_7-DNA of 25×10^6 daltons (Lang, 1970), and assuming that PSTV in urea is formed by two more or less base-paired strands

---

Figure 30. Electron micrographs of PSTV spread in 0.1 M NaCl, 0.01 M EDTA, and 8 M urea at a concentration of 0.6 μg/ml onto a hypophase of 0.015 M ammonium acetate, pH 8. Scale lines = 200 nm. (a) Sample kept at 0°C; (b) Sample heated for 10 minutes at 63°C and then quenched in ice water; (c) Sample after treatment with 10 μg/ml RNase for 1 hour at 25°C. In micrographs (a) and (b), filamentous molecules with a length on the order of 50 nm are recognizable (arrows). These molecules are sensitive to RNase (c). The lengths of heated and unheated molecules differ only slightly, presumably because of rapid renaturation after cooling. Looped strand in (c) is gel impurity. From: Sogo et al. (1973).

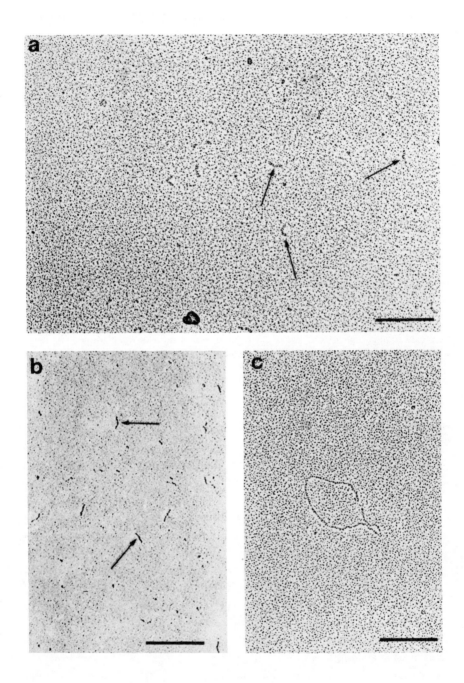

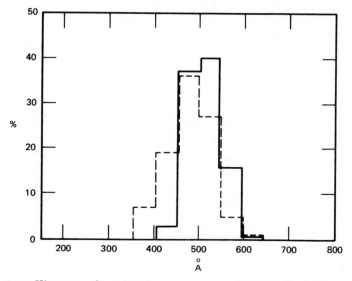

Figure 31. Histograms from length measurements of PSTV (in 8 M urea, 0.1 M NaCl, and 0.01 M EDTA, pH 7) spread at room temperature. (———), Spreading onto 0.015 M ammonium acetate, pH 8; (- - - - -), spreading onto distilled water. The mean length of 542 molecules spread onto ammonium acetate was 51.2 ± 4.1 nm, and that of 575 molecules spread onto water was 50 ± 7.5 nm. From Sogo *et al.* (1973).

(either as a hairpin or a double helix), then one obtains a molecular weight estimate for PSTV of 8.9 × 10⁴ daltons.

In other experiments, mixtures of PSTV and a single-stranded viral RNA were examined and measured. PSTV appeared thicker than this single-stranded viral RNA, and from length comparisons, a molecular weight of 7.9 × 10⁴ daltons was obtained (Sogo *et al.*, 1973).

McClements and Kaesberg (1977) similarly prepared native PSTV for electron microscopy and confirmed the length measurements of Sogo *et al.* (1973) by reporting that under nondenaturing conditions PSTV appears as seemingly double-stranded rods 50 ± 10 nm in length.

Sänger *et al.* (1976), on the other hand, reported that electron microscopy of native CPFV discloses rods or dumbbells with an average particle length of only 35 nm. They concluded that in view of the molecular weight of the viroid (110,000), corresponding to 330 nucleotides, the measured length implies a high degree of base-pairing and condensation by coiling. However, the lengths of native CPFV molecules illustrated (Sänger *et al.*, 1976; Fig. 3) does not agree with the average length of 35 nm reported by the authors; the average length, as compared with the scale line of 100 nm given is barely more than 20 nm. Evidently a contradiction exists and it would be interesting to know which value is the correct one.

Sogo *et al.* (1973) had already attempted to visualize PSTV after denaturation of the RNA molecules with formaldehyde but were not successful. McClements (1975) first achieved this task, and further reports on the visualization of denatured viroid molecules were published by Sänger *et al.* (1976) and McClements and Kaesberg (1977). Because these studies shed light on the question of circularity of viroids, they are described in Section 7.2.2.2.

7.2.1.5 Conclusions from Nucleotide Sequence

The nucleotide sequence of PSTV (*Gross et al.*, 1978) precludes a perfect intramolecular base complementarity within the viroid molecule. Accordingly, the rigid, rodlike native secondary structure of PSTV, as seen in the electron microscope, must be based on a defective rather than a homogeneous RNA helix (Domdey *et al.*, 1978) in conformation of conclusions arrived at earlier by different techniques.

7.2.2 Circular or Linear Ribonucleic Acid

Early experiments with PSTV and CEV indicated that viroids are relatively resistant to attack by exonucleases. These observations led to speculations that viroids may be circular molecules. Definitive evidence for this conjecture was obtained later by electron microscopy of denatured viroids.

7.2.2.1 Exonuclease Resistance

Diener (1970, 1971c) suggested on the basis of resistance of PSTV to exonucleases (see Chapter 4) that PSTV is either a circular molecule or is "masked" at both the 3'- and 5'-terminus in such a manner that exonucleases cannot attack the terminal nucleotide. Because incubation with alkaline phosphatase did not lead to loss of resistance to exonucleases, it follows that this "masking" cannot be due to phosphorylation of the terminal nucleotides.

Semancik and Weathers (1970a) speculated that involvement of a circular structure might explain the existence of two differently sedi-

menting forms of CEV. They conjectured that the sedimentation properties would be markedly influenced by the secondary structure of the
ring as well as either an open or closed configuration, both of which
might be infectious, but to varying degrees. The authors tested this
hypothesis by incubation of CEV preparations with snake venom or
spleen phosphodiesterases, followed by bioassay on a suitable host.
They found that under conditions where tobacco rattle virus RNA
(a linear control RNA) was completely inactivated by either exonuclease,
CEV was still infectious (Semancik and Weathers, 1970a). The resistance
of CEV to inactivation by exonuclease was, however, not complete,
presumably (so the authors surmised) because of endonuclease contamination. The resistance of CEV to exonucleases could also be explained by the lack of free 3'- or 5'-hydroxyl groups at the termini
without the need for invoking a circular structure (Semancik and
Weathers, 1970a).

Additional experiments (Semancik and Weathers, 1972a) indicated
that the resistance of CEV to exonuclease was not sufficient to postulate
a circular structure. The authors concluded that the partial resistance
of CEV might be explained by a delayed inactivation of the double-
stranded regions as compared to the presumed single-stranded form
(Semancik and Weathers, 1972a).

Singh and Clark (1971a) explored the possibility of utilizing the
resistance of PSTV to exonuclease digestion in efforts to purify the
viroid, but came to the conclusion that PSTV is not completely resistant
to attack by snake venom phosphodiesterase.

7.2.2.2 Visualization of Denatured Viroids

McClements (1975) and McClements and Kaesberg (1977) treated purified
PSTV with various concentrations of formamide and prepared the
samples for electron microscopy by spreading onto a water hypophase,
according to the method of Inman and Schnös (1970). With formamide
concentrations of 23%, only seemingly double-stranded rods were found
(see Section 7.2.1.4). Use of higher formamide concentrations in the
spreading mixture led to partial denaturation of PSTV. At a concentration of 45%, undenatured, partially denatured, and fully denatured
molecules were present. As shown in Figure 32, four structures could
be distinguished under these conditions: (1) "double-stranded" rods, (2)

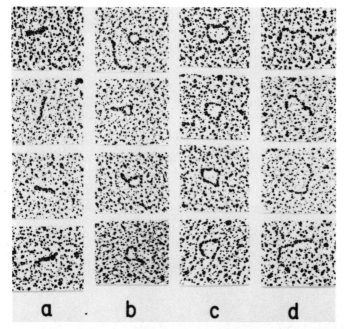

Figure 32. Composite electron micrograph of untreated PSTV. PSTV (approximately 30 μg/ml) was spread. Final formamide concentration was 45%. Spreadings were stained with uranyl acetate and shadowed with Pt. Magnification: *ca. 150,000* ×. (*a*) Native PSTV, double-stranded rods; (*b*) partially denatured hairpin PSTV; (*c*) completely denatured circular PSTV; (*d*) completely denatured linear PSTV. From: McClements and Kaesberg (1977).

partially denatured hairpin molecules, (3) completely denatured circular molecules, and (4) completely denatured linear molecules.

The presence of single-stranded and partially single-stranded molecules suggested to the authors that they were the results of the denaturation of the 50 nm, seemingly double-stranded, rods.

The regularity of the single-stranded molecules was determined by measuring their contour lengths. The mean length of the linear molecules was 90 ± 10 nm, and the mean length (circumference) of the circular molecules was 110 ± 10 nm (McClements and Kaesberg, 1977). Statistical analyses indicated that the length difference between the two types of single-stranded molecules is significant. These lengths are about twice that of the native, seemingly double-stranded rods, further suggesting that both types of single-stranded structures are denatured rods.

In further studies on viroid denaturation, McClements and Kaesberg (1977) showed that at formamide concentrations of 23%, 26%, and 51%, the relative amounts of native PSTV varied from 100 to 64% to 46%, respectively.

To achieve elimination of all secondary structure of PSTV, still higher concentrations of formamide in the spreading mixture could be used. High formamide concentrations, however, degrade spreading quality, and for this reason denaturation was achieved by pretreatment of the RNA with formaldehyde by the method of Boedtker (1971), involving heating of the RNA in 1.1 M formaldehyde for 15 minutes at 63°C, followed by rapid cooling to 0°C. The formylated RNA was then spread at a moderate formamide concentration of 45% (McClements and Kaesberg, 1977).

As shown in Figure 33, only three structures were observed: (1) seem-

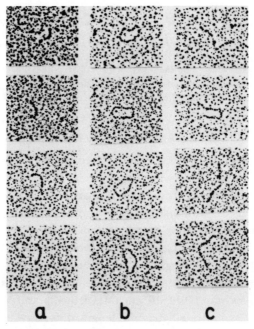

Figure 33. Composite electron micrograph of formaldehyde-treated PSTV. PSTV (approximately 20 µg/ml) was formaldehyde treated, spread, and prepared for microscopy as in Fig. 32. Magnification: *ca. 117,000* ×. (*a*) Native PSTV; (*b*) completely denatured circular PSTV; (*c*) completely denatured linear PSTV. From: McClements and Kaesberg (1977).

ingly double-stranded rods, (2) single-stranded circles, and (3) single-stranded linear molecules. No hairpin structures were detected in any preparations of formylated RNA.

Measurements of formaldehyde-treated molecules also revealed a difference in length between the circular and linear molecules. The mean length of the linear form was 110 ± 10 nm and that of the circular form was 140 ± 10 nm. The distributions of lengths were normal, and by the t test the difference is significant (McClements and Kaesberg, 1977).

The length difference of 30 nm between the linear and circular forms argues against the notion that the circular molecules are linear forms closed by short double-stranded stretches, or that the linear molecules are simply nicked circles.

The complete absence of hairpinlike molecules after formaldehyde treatment of PSTV and the presence, under these conditions, of circular molecules indicate that the circles are covalently closed rather than held together by invisibly short base-paired segments. The authors did not rule out the possibility, however, that the circles might be held together by an unusual, noncovalent type of closure.

Single-stranded, covalently closed circular RNA molecules have not been found previously. These observations constitute strong evidence for the unique molecular structure of viroids. Clearly, preparations of native PSTV consist of a population of two types of seemingly double-stranded RNA molecules that are indistinguishable by electron microscopy: (1) an extensively, but not completely, base-paired linear molecule resembling a hairpin (snapback RNA), and (2) an extensively, but not completely, base-paired circular molecule, also resembling a hairpin, but with covalent closure at both ends.

With two independent preparations of PSTV, about 70% of the molecules were linear ones, about 20% were circular ones, and about 10% remained native even after formaldehyde treatment. Upon denaturation, the hairpinlike linear molecules unfold to become the single-stranded linear RNA molecules depicted by McClements and Kaesberg (1977), whereas the collapsed circular molecules open to become the single-stranded covalently closed circular RNA molecules also illustrated by the same authors.

Denaturation of the linear molecules seems to always start at the closed or loop end of the hairpin and to proceed toward the open end, as evidenced by the fact that partially denatured molecules always exist in the shape of tennis rackets or balloons, but never in the form of "Y"-shaped molecules (McClements and Kaesberg, 1977). The larger susceptibility to denaturation of the loop end of linear PSTV molecules may be the result (so the authors surmised) of a lower proportion of G-C type base pairs, or generally fewer base pairs at that end of the molecule.

Left unresolved by McClements and Kaesberg's (1977) study was the relationship between circular and linear PSTV molecules and the biological significance of either.

In light of McClements and Kaesberg's (1977) results, the two viroid structures could be two distinct RNA components or represent two stages of maturity of a single RNA species. RNA fingerprinting patterns (see Chapter 8) of a mixture of the two components are consistent with a single RNA sequence containing 250 to 350 bases (Dickson *et al.*, 1975), and it seems likely, therefore, that the two structures are two forms of the same RNA (see Chapter 9). The authors speculated that the linear molecules (which evidently are not simply nicked circles) arise from the circular ones by a specific double-stranded cut about 40 base pairs in from one loop of the native circular PSTV molecule with the subsequent loss of the small fragment (McClements and Kaesberg, 1977).

These results and conclusions are in conflict with a report by Sänger *et al.* (1976) in which the authors similarly present evidence (obtained by electron microscopic visualization of denatured viroids) for the presence, in preparations of purified viroids, of single-stranded covalently closed circular RNA molecules existing as highly base-paired rodlike structures. In contrast to the work of McClements (1975) and McClements and Kaesberg (1977), however, Sänger *et al.* (1976) consider viroids to be exclusively circular RNA molecules and linear molecules to be rare (0.5 to 1.0%) and simply representing nicked circles. The authors' electron micrograph depicting denatured viroid reveals, aside from some fully extended circular molecules, a majority of forms resembling "tennis rackets" or "balloons" and some linear molecules. Sänger *et al.* (1976) consider all these structures to be the result of partial renaturation of circular molecules; however, several of the linear molecules are longer than is compatible with a collapsed circle (which can be no longer than one-half the contour length of the fully extended circular molecule). Also, in view of McClements and Kaesberg's observations indicating that denaturation of linear PSTV molecules always begins at the loop end of the hairpin (see above), an unknown number of the racketlike and balloonlike structures shown by Sänger *et al.* (1976) may in fact represent partially denatured (or renatured) linear molecules and not partially collapsed circles. It can be estimated that the proportion of linear molecules in the denatured viroid preparation shown by Sänger *et al.* (1976) was at least 20%, but it could have been as high as 60%. Thus there does not appear to be a basic contradiction between the results of McClements and Kaesberg (1977) and those of Sänger *et al.* (1976).

Evidence for the biological significance of linear (as well as circular) viroid molecules is presented in Chapter 9.

7.3 STABILITY

Since the early days of plant virology, determination of the effects of various physical factors on the integrity of viral agents has been an important criterion for their classification and in establishing relationships with other agents. Before purified pathogens were available, integrity of the agents was judged on the basis of their biological activity. Prominent among these factors was heat, and determination of the "thermal inactivation point" of a viral agent was given high priority in early endeavors.

7.3.1 Effect of Heat

Probably the first effort to establish the "thermal inactivation point" of a viroid was made by Samson (1927), who exposed cores of tissue from PSTV-infected potato tubers to heat in a water bath for 10-minute periods. He found that infectiousness of such cores was destroyed when heated at 60°C. Evidently, in view of more recent work (see below), Samson (1927) did not determine the intrinsic heat stability of PSTV, but rather destruction of a factor necessary for transfer of the viroid from the infected cores to the healthy tubers. Because the temperature at which PSTV appeared to be inactivated closely coincided with that necessary to kill the tuber cores, it appears reasonable to assume that, in Samson's tests, it was the death of the tissue, and not inactivation of PSTV, that prevented infection of the healthy tubers.

Goss (1931) confirmed the results of Samson (1927), but also determined the "thermal death point" of PSTV by the more conventional method of exposing sap expressed from infected tissue to various temperatures, followed by bioassay. With this method, Goss (1931) obtained 30% infection after exposure of sap for 10 minutes to 75°C, but no infection after exposure to 85°C. Thus the "thermal death point" was found to be above 75°C.

Much later, similar results were reported by Singh and Bagnall (1968) for PSTV in tissue extracts. Infectivity was partially lost by exposure of such extracts for 10 minutes to 70°C or 75°C and completely by exposure to 80°C. Heating of total nucleic acid preparations from infected tissue, however, yielded a thermal inactivation "point" that is considerably higher than that obtained with crude tissue extracts.

Exposure of nucleic acid preparations containing PSTV to 90°C did not result in any appreciable loss of infectivity; only after boiling for 5 minutes or autoclaving for 20 minutes at 121°C was infectivity totally lost (Singh and Bagnall, 1968). Evidently, here again, loss of infectivity in a crude tissue extract is not a reflection of the intrinsic heat sensitivity of the viroid, but must be due to other factors. Among these, hydrolysis of PSTV by ribonuclease would appear to be one of the more likely possibilities, but in Singh and Bagnall's (1968) tests, when infectivity of tissue extracts was removed by heating (below that which would inactivate nucleic acid) or by ribonuclease, it could be partially restored by means of phenol treatment. These findings, however, have neither been confirmed nor explained.

Heat inactivation rates in expressed sap similar to those of PSTV were reported by McClean (1935) for TBTV. No loss of infectivity was detected after exposure of sap to 60°C, a definite loss occurred between 60°C and 70°C, and at 80°C inactivation was nearly complete. Some residual infectivity, however, was detectable even after heating to 85°C (McClean, 1935).

With CEV, Roistacher et al. (1969) reported that heating contaminated knife blades for 4 to 8 seconds in a flame only partially inactivated the viroid and that dipping of such knife blades in 95% ethanol, followed by flaming, did not measurably reduce transmission of CEV. The extraordinarily high resistance of CEV to heat inactivation was confirmed by Semancik and Weathers (1970a). The infectivity of ethanol-concentrated high-speed supernatants from tissue extracts was only partially lost by exposing the preparations for 10 minutes to 100°C and phenol-treated preparations did not suffer any loss of infectivity when heated for 10 or even 20 minutes at 100°C. Further experiments showed that infectivity was not affected by temperatures up to 110°C, but decreased in a linear fashion to a thermal death "point" of about 140°C (Semancik and Weathers, 1972a).

In common with other viroids, CSV possesses an exceptional heat tolerance. Brierley (1952) discovered that this viroid readily survives boiling for 10 minutes in a tissue extract, but Keller (1953) determined in further tests that the thermal inactivation "point" of CSV is between 96°C and 110°C. Hollings and Stone (1973) reported that CSV withstands 10 minutes at 90°C to 100°C.

Undoubtedly, this exceptional heat tolerance of viroids is responsible for the fact that procedures useful for the *in vivo* eradication of a number of plant viruses have had only limited success when applied to viroids. Thus thermotherapy of PSTV-infected potato plants, followed by axillary bud culture, led to eradication of PSTV in only a few of the many plantlets developed (Stace-Smith and Mellor, 1970) and heat-treatment of CSV-infected chrysanthemum plants, followed by meristem-tip culture, similarly failed to eradicate the viroid, except in a very few cases (Hollings and Stone, 1970).

7.3.2 Effect of Cold

Keller (1953) reported that CSV tolerates freezing and thawing both in extracted juice and in leaf tissue, and that the viroid can be stored frozen for several weeks without undergoing appreciable inactivation.

Hunter and Rich (1964b) compared the infectivity levels of PSTV in tomato leaf extracts prepared by three methods. They found that freezing infected leaves in liquid nitrogen (− 196°C), grinding the frozen tissue to a fine powder, and adding this powder to buffer consistently resulted in higher infectivity levels and more rapid symptom formation than either grinding fresh leaves in buffer or adding expressed juice to buffer. Evidently, PSTV tolerates freezing and thawing.

Not surprisingly, in view of these results, PSTV can be freeze-dried and stored for at least 6 years at room temperature, without significant loss of infectivity (Singh and Finnie, 1977).

7.3.3 Effects of Other Physical Factors

CSV retains infectivity in air-dried, infected leaves for periods of storage at room temperature of 2 to 4 weeks in some trials, at least 8 weeks in others (Brierley, 1952), and at least 100 weeks in still other trials (Keller, 1953).

On contaminated knife blades, CEV remains infectious for at least 8 days (Allen, 1968).

Viroids are relatively tolerant to shearing forces as indicated by results of Hollings and Stone (1973), who subjected CSV in buffer extracts to sonication for 5 or 15 minutes and observed only partial loss of infectivity.

8. CHEMICAL PROPERTIES OF VIROIDS

At an early stage of viroid research it had become evident that RNA was an essential constituent of these pathogens, but only later work made it apparent that no other constituents were associated with the infectious RNA or necessary for its biological activity.

8.1 CONSTITUENTS OTHER THAN RIBONUCLEIC ACID

As discussed in Chapter 4, all available evidence indicated that no virionlike nucleoprotein particles were present in viroid-infected tissue, but the possibility remained that a small protein might be associated with the viroid RNA and that this protein might be of biological significance.

Incubation of extracts from PSTV-infected tissue with Pronase revealed, however, that this treatment neither affects the infectivity titer nor the sedimentation properties of the infectious particles (Diener, 1971a). Similar results were obtained by Sänger (1972) for CEV. No loss of infectivity was observed when CEV-containing preparations were incubated with trypsin, chymotrypsin, or Pronase. Later, electron microscopy of denatured viroids showed that treatment with Pronase or proteinase K did not alter any of the structures observed before such treatment (Sänger et al., 1976). Thus no residual protein components appear to be associated with purified viroids, and the circular structures are not held together by a protein linkage.

Semancik et al. (1975) analyzed preparations of CEV for RNA, DNA, and protein content by the orcinol, diphenylamine, and Folin tests, respectively. No significant levels of either DNA or protein could be detected, and the RNA content as determined by the orcinol test was within 80%–85% agreement with the value obtained by using an extinction coefficient of $E\frac{0.1}{260} = 20$.

Early experiments with PSTV had shown that treatment of infectious extracts with DNase had no effect on the infectivity titer of the preparations (Diener and Raymer, 1967), but such treatment appeared to lead to a lowering of the sedimentation coefficient of the infectious material (Diener and Raymer, 1969). Later work, however, did not confirm this observation (Diener, 1971a). Possibly, the early observations were due to incomplete detachment of PSTV from host DNA (see Chapter 9).

Singh *et al.* (1975a), on the other hand, reported that PSTV from infected *Scopolia sinensis* plants contain an infectious component that consists of RNA associated with DNA. Existence of the DNA was established by its separation by high pressure liquid chromatography and by DNase digestion followed by gel electrophoresis. Singh *et al.* (1975a) implicated the DNA-associated RNA in a role in viroid replication. These observations, however, have not been confirmed by other workers with other viroids or with the same viroid (PSTV) isolated from different hosts.

Present evidence thus indicates that the only component necessary for the expression of the biological activity of viroids is RNA and that neither RNA-associated DNA nor protein is required for the initiation of viroid replication.

8.2 ONE OR SEVERAL RIBONUCLEIC ACID SPECIES?

Sedimentation and gel electrophoretic analyses of infectious extracts from PSTV-infected tissue, followed by determination of infectivity distribution, revealed several areas in density gradients or polyacrylamide gels with significant levels of infectivity (Diener and Raymer, 1969; Diener, 1971b). This paucidisperse distribution of PSTV was not an artifact of the analytical methods used, but was due to the existence of infectious RNA molecules with different molecular weights (Diener, 1971b). The higher molecular weight species sedimented as would be expected if they were dimers, trimers, and tetramers of a minimal infectious unit. Thus the hypothesis was advanced that PSTV occurs in extracts in the form of aggregates of varying sizes which, under certain conditions (such as electrophoresis in gels of small pore size), may disaggregate (Diener, 1971b). Indeed, electrophoresis of PSTV in 20% polyacrylamide gels resulted in a single infectious component (Diener and Smith, 1971) and, because no infectious RNA could be detected at the origin or in the buffer immediately above the gel surface, all infectious RNA was evidently able to enter the gel. These observations strengthen the hypothesis of self-aggregation of PSTV and indicate that, under appropriate conditions, only one band containing infectious RNA is present. According to later work, this band of infectious RNA contains only one molecular weight species (see Section 8.3.3). Consequently, despite the early observations of several zones of infectious RNA in density gradients and polyacrylamide gels, only one molecular species of PSTV appears to exist. Supportive evidence for this contention was obtained in biological experiments (see Chapter 9).

These conclusions are compatible with those obtained with other viroids that have been intensively studied. Thus Semancik and Weathers (1972b) considered CEV to consist of one species of infectious low molecular weight RNA, and later evidence obtained by RNA fingerprinting helped to substantiate this conclusion (Dickson *et al.*, 1976; see Section 8.3.3).

The work of Sänger *et al.* (1976) and Henco *et al.* (1977) similarly implies that PSTV, CEV, and CPFV each consist of a single, characteristic molecular species of RNA.

Singh and Clark (1973) reported that when PSTV isolated from *Scopolia* plants was subjected to preparative gel electrophoresis, the infectious RNA was confined to a single, well-defined, and sharp peak. Further analysis of material from this peak on analytical gels indicated that the infectious RNA consisted of a single component (Singh and Clark, 1973).

Contrary to these findings, Singh *et al.* (1974) described the isolation, from infected *Scopolia,* of multiple forms of PSTV by gel filtration in Bio-Gel P-200 columns. The smallest infectious form was reported to be in the size range of tRNA, as estimated by its position in the gel filtration profile. Later, Singh *et al.* (1976) used additional techniques to separate multiple infectious forms of PSTV isolated from *Scopolia* plants. Subjecting of PSTV preparations to either polyacrylamide gel electrophoresis in 10% gels, to reverse phase chromatography, or to high pressure liquid chromatography resulted, with each method, in separation of the infectious RNA into three zones. Two of these components were further purified, hydrolyzed with pancreatic or T_1 ribonucleases, and labeled with $[\gamma^{32}P]ATP$ by polynucleotide kinase (Singh et al., 1975a, b). Two-dimensional RNA fingerprinting indicated that the two forms of PSTV were distinct and dissimilar RNA species (Singh et al., 1975a, b).

Evidently, these results are in conflict with those of other investigators. Conceivably, PSTV isolated from *Scopolia sinensis* has different properties than PSTV isolated from other hosts, but in light of recent work which indicates that the primary viroid sequence is maintained irrespective of the host in which replication occurs (see Chapter 9), this explanation appears implausible. Also, Singh et al. (1976) state that the difference in mobility of the infectious RNA and in the number of components has also been observed in tomato and potato. Thus the observations by Singh and coworkers of multiple PSTV components do not appear to be due to host-dependent effects.

In 3% polyacrylamide gels, some PSTV infectivity has regularly been found in the tRNA region (Diener, 1971b). This led to the hypothesis that the smallest infectious entity had a molecular weight similar to that of tRNA. In 20% polyacrylamide gels, however, only traces or no infectivity was found in the tRNA region (Diener and Smith, 1971) and the earlier observations were considered to be due to artifacts of the analytical system.

After the observations of Singh and coworkers had become known, the significance of PSTV infectivity in the tRNA region of gels was reinvestigated by collecting infectious RNA that had migrated into the tRNA region from several gels, extracting the RNA and subjecting it to a second cycle of gel electrophoresis. Determination of the infectivity distribution across the gel by bioassay on tomato revealed that almost all of the RNA now migrated to the regular position of PSTV, and not to the tRNA region (Diener, unpublished results). Thus no evidence was found for multiple PSTV components. On the contrary, these results strengthen the previously advanced hypothesis of separation artifacts.

8.3   CHEMICAL COMPOSITION

So far, the nucleotide sequence has been reported for only one viroid, PSTV (Gross et al., 1978). Thus the chemical composition of other viroids is known only partially.

8.3.1   Nucleotide Ratios

Semancik *et al.* (1975) determined the nucleotide composition of CEV and Niblett *et al.* (1976) reported that of PSTV. As shown in Table 25, the values for the two viroids are very similar. With either viroid, A/U and G/C ratios approach unity and the G + C ratio is between 55 and 58 mole %.

In addition to the four major nucleotides, Niblett *et al.* (1976) detected a minor component comprising about 0.3% of the RNA. The identity of this component has not yet been determined.

The nucleotide ratio of PSTV, as determined from its nucleotide sequence, is consistent with earlier determinations (Table 25).

8.3.2   End-Group Analyses

Singh and Clark (1973a) subjected PSTV to the Randerath *et al.* (1972) procedure and reported that the 3'-terminal of PSTV was adenosine. Sänger *et al.* (1976), on the other hand, found that CPFV was not ^{3}H-labeled upon metaperiodate oxidation and [^{3}H]borohydride reduction, suggesting that the viroid lacked a free 3'-terminal. Also, analysis in a 10% polyacrylamide gel showed that viroids were not phosphorylated at all by 5'-polynucleotide kinase, suggesting the absence of an accessible 5'-terminal. Radioactivity, however, was found at a position corresponding to RNA that is moving significantly more slowly than viroids (Sänger *et al.*, 1976). This RNA, according to the authors, could be either "nicked viroid RNA" or a contaminating host RNA species.

Table 25.   Nucleotide Composition of Viroids[a]

Viroid	Method	Reference[b]	AMP	GMP	CMP	UMP	A/U	G/C	%G+C
CEV	A_{260}	1	21.5	28.8	29.4	19.9	1.08	0.98	58.2
CEV	^{32}P	1	21.3	27.3	28.3	23.0	0.93	0.96	55.6
PSTV	Randerath	2	21.7	28.9	28.3	20.9	1.04	1.02	57.2
PSTV	Sequence	3	20.3	28.1	30.1	21.4	0.95	0.93	58.2

[a] Expressed in mole %.
[b] References: 1 = Semancik *et al.* (1975); 2 = Niblett *et al.* (1976); 3 = Gross *et al.* (1978).

8.3.3   Oligonucleotide Patterns

Dickson *et al.* (1975) labeled PSTV purified from tomato and CEV purified from *Gynura in vitro* with ^{125}I by using the Commerford reaction (Commerford, 1971; Prensky, 1975), digested the labeled viroids with ribonuclease A or T_1, and separated the oligonucleotides by two-dimensional fingerprinting techniques (Brownlee and Sanger, 1969). Autoradiography resulted in the fingerprint patterns illustrated in Figures 34 and 35. These

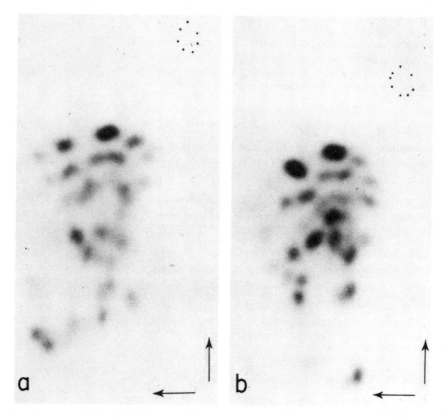

Figure 34.   Ribonuclease T_1 fingerprints of ^{125}I-labeled PSTV and CEV. About 1 × 10[6] dpm of ^{125}I-labeled RNA were mixed with 10 μg of bacteriophage f2 RNA and digested with 2 μg of RNase T_1 in 2 μl of 0.01 M Tris-HCl (pH 7.5) — lm$M$ EDTA for 40 minutes at 37°C. In the configuration shown here, the first dimension of fingerprinting was from right to left and the second was from bottom to top as shown by the arrows. (*A*) PSTV iodinated *in vitro* to a specific activity of 12 × 10[6] dpm per μg. (*B*) CEV iodinated *in vitro* to a specific activity of 18 × 10[6] dpm per μg. From: Dickson *et al.* (1975).

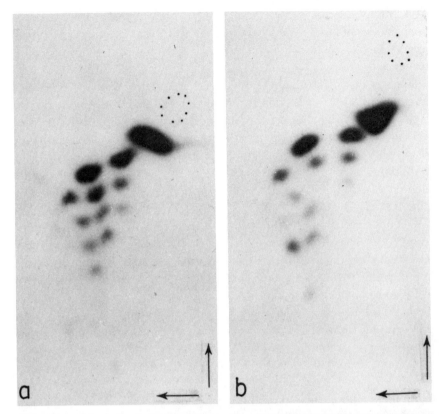

Figure 35. Pancreatic ribonuclease fingerprints of ¹²⁵I-labeled PSTV and CEV. About 1 × 10⁶ dpm of ¹²⁵I-labeled RNA were mixed with 10μg of bacteriophage f2 RNA and digested with 2 μg of pancreatic ribonuclease in 2 μml of 0.01 *M* Tris-HCl (pH 7.5) − 1m*M* EDTA for 30 minutes at 37°C. Two-dimensional fingerprinting analyses were then carried out. The origin is at the lower right; the electrophoretic first dimension was from right to left, and the second dimension (homochromatography) was from the bottom of the picture to the top. (*A*) PSTV iodinated *in vitro* to a specific activity of 12 × 10⁶ dpm per μg. (*B*) CEV iodinated *in vitro* to a specific activity of 18 × 10⁶ dpm per μg.

patterns indicate that each of these viroids has a complexity that is compatible with the size estimate of 250 to 350 nucleotides (Dickson *et al.*, 1975). These results also clearly demonstrate that PSTV and CEV do not have the same primary sequence, a result that contradicts conclusions drawn by some workers from biological experiments which suggested that the two viroids were independent isolates of the same pathogen (Semancik and Weathers, 1972c; Singh and Clark, 1973b; Semancik *et al.*, 1973a).

Also, the fingerprint patterns indicate that each of these viroid preparations consists of a single molecular species of RNA and not of a mixture of several RNA species of about equal length but different nucleotide sequence.

Singh *et al.* (1976) obtained oligonucleotide patterns of their so-called fraction II PSTV (see Section 8.2) by digesting this fraction with ribonuclease T_1 or A, labeling of the oligonucleotides with $[\gamma\text{-}^{32}P]ATP$ using polynucleotide kinase, and separating the labeled oligonucleotides by the electrophoresis-homochromatography procedure of Brownlee *et al.* (1968). Autoradiography disclosed approximately 20 oligonucleotide spots after ribonuclease A digestion and about 30 spots after ribonuclase T_1 digestion (Singh *et al.*, 1976).

Oligonucleotide fingerprints of PSTV from tomato, CEV from *Gynura*, and CSV from cineraria were obtained by Gross *et al.* (1977). Each purified viroid was digested with either ribonuclease A or T_1, and the resulting oligonucleotides were then labeled at their 5'-terminals with $[\gamma\text{-}^{32}P]$-ATP and polynucleotide kinase. Two-dimensional fingerprints (obtained by high voltage electrophoresis-homochromatography, followed by autoradiography) showed that the number of oligonucleotides and their intensity in the autoradiograms were, in each case, consistent with a viroid of about 320 to 380 nucleotides and that the fingerprints of the three viroids analyzed contained significant and characteristic differences in both their ribonuclease A and T_1 patterns (Gross *et al.*, 1977). These observations demonstrate the individuality of the three viroids investigated and therefore confirm the conclusions reached by Dickson *et al.* (1975).

One puzzling aspect of these studies concerns the fingerprint patterns reported by Singh *et al.* (1976) which, as has been noted by Gross *et al.* (1977), are completely different from the patterns illustrated by the latter workers and contain far fewer oligonucleotides. In part, this discrepancy may be due to the removal before fingerprinting of dinucleotides in the studies of Singh *et al.* (1976) but not in those of Gross *et al.* (1977).

8.3.4   Nucleotide Sequence Determination

Domdey *et al.* (1978) and Gross *et al.* (1978) brought these studies to their logical conclusion by determining the complete nucleotide sequence of PSTV. The viroid was completely digested with RNase T_1 and RNase A, and the resulting oligonucleotides were sequenced using 5'-terminal ^{32}P-

labeling with $[\gamma\text{-}^{32}P]ATP$ and T_4 polynucleotide kinase, fingerprinting, and controlled nuclease P_1 digestion. No nucleotides modified in their 5' positions were detected. The results indicate that PSTV consists of about 359 nucleotides and contains an unusual stretch of 18 purines, mainly adenosines. Interestingly, the sequence does not contain an AUG initiation triplet (Domdey et al., 1978). From the established nucleotide sequence a most likely secondary structure has been deduced (see Section 7.2.1.5).

8.3.5 Absence of Poly A and Poly C Sequences

Semancik (1974) searched for polyadenylic acid sequences in CEV by attempting to hybridize the viroid with [^{3}H]polyuridylic acid, followed by determination of RNase-resistant radioactivity. No evidence for the presence of poly A sequences was obtained (Semancik, 1974).

Hadidi et al. (1977) synthesized DNA transcripts of PSTV using DNA polymerase I from *Escherichia coli* and observed that DNA synthesis did not take place when d[^{3}H]TTP was used as the sole precursor in the presence of PSTV and $(dT)_{10}$ primer, indicating that poly A stretches were absent in PSTV. Poly C stretches also are absent in the viroid, because DNA synthesis was not observed when d[^{3}H]GTP was the sole precursor in a reaction with PSTV and $(dG)_{12\text{-}18}$ primer (Hadidi et al., 1977).

9. VIROID-HOST INTERACTIONS

In this chapter, the events that occur when viroids are introduced into susceptible host cells are discussed. As will be seen, many gaps in our knowledge of these processes still exist. Little is known, for example, of the molecular processes that lead to replication of viroids, and even less of the mechanisms that result in symptom formation. Much research is proceeding in these areas and undoubtedly the mode of viroid replication will, to a large extent, be known in the near future. Less certain are the prospects for elucidation of the mechanisms of pathogenesis, but it must be remembered that little is known, at the molecular level, about analogous processes operative with conventional plant viruses. Because of their extremely limited information content, viroids appear to constitute a far superior model system than conventional viruses for the elucidation of these important mechanisms. Most of the symptom types observed with plant viruses occur also with viroids; thus any knowledge gained with viroid-host systems may be directly applicable to plant virus-host systems.

9.1 INFECTIVITY OF VIROIDS

9.1.1 Specific Infectivity

Mechanical inoculation of partially purified PSTV preparations into tomato plants showed that the viroid has an unusually high specific infectivity (Diener and Raymer, 1967). PSTV concentrates with infectivity dilution end points of 10^{-6} or 10^{-7} gave no recognizable ultraviolet light absorbing peaks in sucrose gradients, columns of methylated serum albumin, or polyacrylamide gels. Furthermore, these early experiments revealed that fractions eluted from cellulose columns that contained only 4 to 5 μg of nucleic acid per ml had infectivity dilution end points of 10^{-3} or 10^{-4} (Diener and Raymer, 1967). Thus solutions containing as little as 5×10^{-4} μg of total nucleic acid per ml, most of which undoubtedly was host nucleic acid, were infectious. Infectivity at such low concentrations has not been observed with any plant viral RNAs. Inoculation of plants with purified viroids confirmed the extremely high specific infectivity of these pathogens. Sänger *et al.* (1976), for example, found that inoculation of tomato plants with 50 to 100 molecules per plant resulted in 10% of the plants becoming infected.

In view of the relative insensitivity of mechanical inoculation techniques in plants, in which the ratio of virus particles (or nucleic acid molecules) to infective units may be 10^6 or larger (Steere, 1955),[1] the specific infectivity of viroids is astounding. Conceivably, in contrast to conventional viruses and viral nucleic acids, viroids might be able to enter host cells without the necessity of wounding. Experiments in which PSTV preparations were sprayed onto the surfaces of unwounded leaves demonstrated, however, that this is not the case (Diener, unpublished). As with conventional plant viruses and viral nucleic acids, wounding of cells is essential for infection to occur.

9.1.2 Viroid Concentration and Infectivity

Bioassay of serial 10-fold dilutions of PSTV preparations on a systemic host, tomato, seemed to indicate that only one type of PSTV molecule was necessary for initiation of infection (Raymer and Diener, 1969), but only with the use of a local lesion host, *Scopolia sinensis* (Singh, 1971), could the slope of the dilution curve be determined more accurately. In Figure 36, the relationship of PSTV concentration to lesion numbers is compared with that of a typical single-component plant virus (tobacco necrosis virus) and with that of a multicomponent virus (alfalfa mosaic virus). Evidently, the dilution curve of PSTV is of the single-hit type, and does not indicate the necessity of two or more types of RNA molecules to be present at each infection site. As discussed in Chapter 8, this conclusion later received ample support from fingerprint analyses of viroid preparations which indicated that each viroid consists of a characteristic single species of low molecular weight RNA.

[1] Fraenkel-Conrat *et al.* (1964), however, presented evidence that with tobacco mosaic virus, at least, the infection process may be more efficient than is generally surmised.

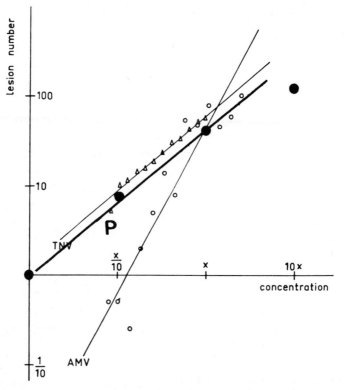

Figure 36. Local lesion response to inoculation with various concentrations of a typical single-component virus (tobacco necrosis virus, TNV), a covirus system (alfalfa mosaic virus, AMV), and PSTV (P). ·TNV and AMV data from van Vloten-Doting *et al.* (1968); PSTV data from Singh (1971).

9.1.3   Infectivity of Circular and Linear Forms

Electron microscopy of denatured viroids reveals two types of molecules: covalently closed circular ones and linear ones (see Chapter 7). It was of obvious interest to assess the significance of the two types of molecules by determining whether one or the other or both possess biological activity.

Owens *et al.* (1977) developed a method for the separation of linear from circular molecules that is based on electrophoresis of formamide-denatured PSTV in polyacrylamide gels equilibrated with 50% formamide and 6 *M* urea. Under these denaturing conditions, PSTV separates into two well-separated bands (Fig. 37). Electron microscopy of unfractionated, denatured PSTV revealed about 18% circular and 82% linear molecules (Fig.

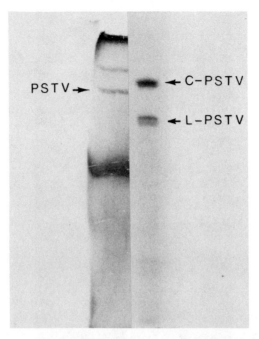

Figure 37.   Purification of circular and linear forms of PSTV. The gel at the left is a strip cut from the edge of a preparative 20% polyacrylamide/0.5% bis-acrylamide slab used for the routine purification of native PSTV. In addition to PSTV, two host RNA species, 5 S ribosomal RNA and 9 S RNA (function unkown), are also visible. The gel at right is an electrophoretic analysis of purified PSTV (7 μg) in a 5% acrylamide/0.125% bis-acrylamide slab containing 50% formamide/6M urea. The upper band (C-PSTV) contains predominantly circular molecules; the lower band (L-PSTV) contains only linear molecules. The trace of 5 S RNA contaminating the native PSTV preparation is visible near the bottom of the gel. From: Owens *et al.* (1977).

38*a*); electron microscopy of RNA recovered from each band showed that the lower band contained almost exclusively linear PSTV molecules (Fig. 38*c*), whereas the upper band contained a majority of circular molecules (70%) but also some linear ones (Fig. 38*b*). Evidently, the appearance of two bands during electrophoresis of denatured PSTV results from the separation of linear from circular molecules. This separation, however, is incomplete. Essentially all circular molecules are contained in the upper band, but linear molecules are detectable in both bands. It is possible that during gel electrophoresis some linear molecules are trapped among the circular molecules. A similar phenomenon has been observed during purification of ^{32}P-labeled PSTV in which, after one cycle of gel electro-

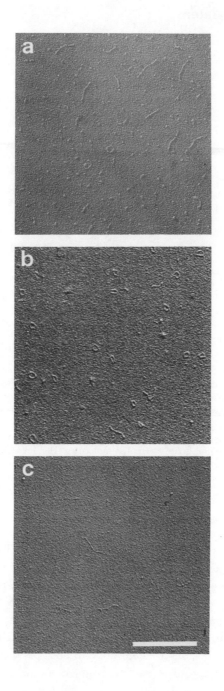

phoresis, PSTV was shown to be contaminated with some 5 S RNA (Hadidi and Diener, 1977). Alternatively, one may conjecture that the upper band originally contained solely circular PSTV molecules but that some of these molecules broke during the process of spreading for electron microscopy, at which time they undoubtedly were subjected to strong shearing forces (Owens et al., 1977).

Bioassay of individual slices taken from the region of the gels containing the two PSTV bands gave the results shown in Figure 39. Two clearly separated regions of infectivity were apparent and these regions coincided with the positions of the two PSTV bands (Owens et al., 1977).

The association of infectivity with the lower band (Fig. 39) and the lack of circular molecules in this band unequivocally demonstrate that the linear molecules of PSTV are infectious. Because of the contamination of the upper band with linear molecules, the case for the infectious nature of the circular molecules is somewhat equivocal as it depends on the tacit assumption that the specific infectivity of linear molecules in the upper band is equal to that of linear molecules in the lower band.

It appears unlikely that randomly nicked circular molecules would retain biological activity. Thus, if the linear molecules originate from circular ones, nicking at a specific site appears likely.

The possibility that randomly nicked molecules might be repaired and reformed into circular molecules after inoculation by a host RNA ligase cannot, however, be excluded, particularly because the very existence of covalently linked circular RNA molecules strongly suggests that such an enzyme exists in host plants of viroids.

9.1.4 Inhibitors of Infection

In contrast to the many known inhibitors of infection with conventional plant viruses, very few studies have been made with viroids.

Singh et al. (1975c) reported that piperonyl butoxide, the active ingredient of certain commercial insecticide formulations, is a potent inhibitor of PSTV lesion formation in Scopolia sinensis, provided the compound is

---

Figure 38. Electron microscopy of PSTV. Unfractionated PSTV and RNA extracted from the two major zones visible after gel electrophoresis of PSTV in the presence of formamide and urea were prepared for electron microscopy. (a) Unfractionated PSTV; (b) RNA extracted from the upper (slower migrating) band in formamide/urea electrophoresis; (c) RNA extracted from the lower (faster migrating) band in formamide/urea electrophoresis. Bar represents 400 nm. From: Owens et al. (1977).

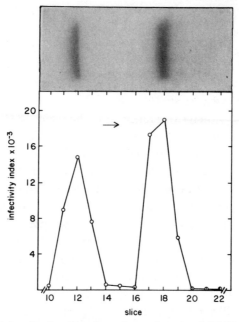

Figure 39. Infectivity distribution after electrophoresis of PSTV under denaturing conditions. Two samples of purified PSTV (7 μg each) were denatured and electrophoresed in the presence of formamide and urea. After electrophoresis, one sample was stained to locate the two PSTV-specific bands (upper). The corresponding area of the second sample was assayed for infectivity (lower); three plants were inoculated with each dilution. The arrow marks the direction of electrophoretic migration. From: Owens *et al.* (1977).

sprayed onto the leaves before inoculation with PSTV or no later than 4 days after inoculation. Gel electrophoretic analysis of low molecular weight RNAs isolated from treated plants and infectivity assays indicate that piperonyl butoxide inhibits the synthesis of PSTV (Singh *et al.*, 1975c).

In a later study, Singh (1977) reported that piperonyl butoxide also inhibits infection of potato by PSTV, but in an experiment designed to simulate field conditions some infection with PSTV occurred despite weekly spraying with the compound.

9.2 VIROID FUNCTIONS IN CELL

Once a viroid enters a susceptible cell, it must be able to trigger biochemical mechanisms that eventually lead to viroid replication and in some hosts to symptom formation. What are these mechanisms?

9.2.1   Messenger Ribonucleic Acid Activity

In analogy with known *in vivo* functions of conventional viral nucleic acids, one could postulate that the viroid might act as a messenger RNA and that the translated polypeptide might be instrumental in initiating the processses known to take place in viroid-infected plants. In view of the small size of viroids, however, such a polypeptide would, by necessity, have to be very small. A protein about 10 to 15,000 daltons, corresponding to 100 to 130 amino acids, would represent the total coding capacity of viroids.

9.2.1.1   In vitro m-Ribonucleic Acid Activity

The question as to whether viroids can act as messenger RNAs in several cell-free protein synthesizing systems was first investigated by Davies *et al.* (1974) with PSTV.

Extracts from wheat embryo, commercial wheat germ, *Escherichia coli*, and *Bacillus stearothermophilus*, which are known to translate RNAs from viruses that do not infect these organisms, were found to be unable to synthesize any specific proteins directed by purified PSTV. Amino acid incorporation was not stimulated much above the control without PSTV. This stimulation was less than 1% of the activity due to $Q\beta$ and brome mosaic virus (BMV) RNAs, which can be translated into specific products under the same conditions.

In the wheat germ system, the stimulation due to 5 μg of PSTV was 2.35% of the BMV RNA control, equivalent to only a 1.3-fold stimulation above the control without RNA. It was not increased by adding more PSTV (Davies *et al.*, 1974). This small stimulation, however, resulted in more than 5×10^3 cpm above the control without RNA, and this amount would be detectable on polyacrylamide gels if incorporated into a definite product. No such product was found, whereas the controls for BMV and $Q\beta$ produced the normal *in vitro* products.

Because PSTV is known to have a compact structure, its ability as a messenger RNA was also tested in *B. stearothermophilus* extracts at 60°C, at which temperature its structure should have been opened up sufficiently to allow ribosome attachment (see Section 7.2.1.3). Even under these conditions, however, no evidence of amino acid incorporation or translation activity was detected (Davis *et al.*, 1974).

Similar experiments were performed with CEV (Hall *et al.*, 1974) with identical results. Four cell-free protein synthesizing systems were tested:

extracts from *E. coli, Pseudomonas aeruginosa, B. stearothermophilus,* and wheat. With none of these could any recognizable products be identified as a result of incubation in the presence of CEV, whereas incubation with control RNAs led to formation of the expected products (Hall *et al.,* 1974).

More recently, Semancik *et al.* (1977) injected CEV into oocytes of *Xenopus laevis* under conditions that led to translation of BMV RNA and rabbit hemoglobin mRNA. The viroid was neither translated nor did it interfere with protein synthesis in the oocytes. These results were not altered when *in vitro* adenylated CEV or total nucleic acid preparations from CEV-infected *Gynura* were injected into oocytes.

The inactivity of viroids as mRNAs in several cell-free protein-synthesizing systems and their inactivity in a simulated *in vivo* system, using *Xenopus* oocytes, strongly suggest that viroids are not translated and that, therefore, their interference with host metabolism does not appear to be mediated via a viroid-specified protein.

9.2.1.2   In vivo *m-Ribonucleic Acid Activity*

It is possible, however, that viroids may be transcribed by preexisting host enzymes into complementary strands and that these then might act as messenger RNAs. If so, a viroid-specific protein should be detectable in infected plants.

Zaitlin and Hariharasubramanian (1972) studied the relative incorporation of leucine-^{3}H or ^{14}C into proteins of PSTV-infected and uninfected tomato leaf tissue to detect viroid-mediated or viroid-stimulated protein synthesis. Proteins from subcellular fractions of PSTV-infected tissue were analyzed by electrophoresis in SDS-containing polyacrylamide gels. No proteins were found in PSTV-infected tissue that could not also be detected in extracts from uninfected leaves; yet, in control experiments, leucine-^{3}H-incorporation into coat protein of the defective tobacco mosaic virus strain PM21 could readily be detected with these techniques (Zaitlin and Hariharasubramanian, 1972). In a "ribosomal supernatant" fraction, however, incorporation of leucine into two high molecular weight proteins was stimulated in PSTV-infected as compared with uninfected tissue. No such stimulation, on the other hand, was found in a "nuclear" fraction.

Interestingly, the two proteins, whose synthesis was stimulated by infection with PSTV, appear to have similar molecular weights, as do two proteins stimulated in tobacco mosaic virus-infected tobacco leaves;

namely, 1.55×10^5 and 1.95×10^5 daltons. Because stimulation of these proteins could not be detected in the "nuclear" fraction, where most of the infectious PSTV is known to be located (see Section 9.3), and because PSTV could not contain sufficient coding capacity to code for high molecular weight proteins such as were observed, the authors concluded that synthesis of these proteins is stimulated by the PSTV infection, but that they are not coded for by the viroid (Zaitlin and Hariharasubramanian, 1972).

In similar studies with protein extracts of CEV-infected and healthy *Gynura* tissue, as well as with lysed protoplasts, Conejero and Semancik (1977) were unable to detect qualitative changes in the protein composition of infected, as compared with healthy, tissue. Quantitative differences, however, were noted, most markedly in postribosomal preparations, in which two proteins with molecular weights of about 15,000 and 18,000 daltons accumulated in infected plants. Because comparable samples from healthy tissue never were completely devoid of detectable amounts of protein in the same regions, the authors considered it unlikely that these two proteins were direct translation products of CEV, but that they rather reflected a quantitative alteration in host-specified proteins.

In conclusion, no evidence exists to indicate that viroids or their transcription products ever act as messenger RNAs either *in vitro* or *in vivo*.

9.2.2 Transfer Ribonucleic Acid Activity

Several RNA components from small viruses are capable of binding specific amino acids in a transfer RNA-like manner (Yot *et al.*, 1970; Öberg and Philipson, 1972; and others). To determine whether viroids can serve as amino acid acceptors, Hall *et al.* (1974) performed aminoacylation reactions with CEV, BMV RNA, tobacco mosaic virus RNA, and genuine tRNA. They found that, in contrast to the control RNAs which readily bound specific amino acids, CEV did not accept any amino acids.

Apparently then, viroid-host interactions cannot be explained by postulating a tRNA-like function for the viroid.

9.3 SUBCELLULAR LOCATION

When PSTV-infected leaves were extracted by the method of Kühl (1964), which maintains the integrity of cell nuclei, infectivity distribution in the various fractions was as shown in Table 26 (Diener, 1971a).

Table 26. Distribution of PSTV Infectivity in Subcellular Fractions[a]

Fraction	Infectivity Indices[b]	
	Experiment 1	Experiment 2
Tissue debris (pulp in cloth)[c]	71	45
1st supernatant after 10 minutes at 350 g		
(soluble, ribosomal, mitochondrial fraction)	0	0
2nd supernatant	0	4
3rd supernatant	4	0
4th supernatant	NT	12
Crude nuclear fraction (nuclei and		
chloroplasts)	103	77
Upper green-layer band (chloroplasts)[d]	0	0
Lower green-layer band (chloroplasts)[d]	0	6
Purified nuclear fraction[d]	106	60
Nuclear fraction after treatment with		
Triton X-100[e]	90	NT

From: Diener (1971a).

[a] Prepared by the method of Kühl (1964).

[b] Preparations assayed undiluted and diluted 10^{-1}, 10^{-2}, 10^{-3}, and 10^{-4} in 0.02 M phosphate buffer, pH 7. NT = not tested.

[c] Suspended in 0.5 M K_2HPO_4.

[d] Prepared by density-gradient centrifugation (Tewari and Wildman, 1966).

[e] Purified nuclei treated with 0.1% Triton X-100 for 1 hour at 0°C, followed by pelleting of the nuclei and resuspension in isolation medium (Spencer and Wildman, 1964).

Considerable infectivity remained associated with the tissue debris. The supernatant solutions after centrifugation for 10 minutes at 350 g were devoid of infectivity. Thus, no infectivity was associated with ribosomes or mitochondria or was present in free form in the cytoplasm. Supernatant solutions, arising from subsequent centrifugations, were similarly devoid or almost devoid of infectiviy. The crude nuclear fractions, on the other hand, were highly infectious.

Since these fractions were contaminated with many chloroplasts, the nuclei were further purified by density-gradient centrifugation (Tewari and Wildman, 1966). The resulting chloroplast-containing bands were essentially noninfectious, whereas the purified nuclear fractions were highly infectious (Table 26).

After treatment of the purified nuclear fractions with Triton X-100, no chloroplasts were detectable. Such preparations contained large numbers of normal-appearing nuclei as well as some crystalline material. Treatment with Triton X-100 had little effect on infectivity (Table 26).

The question arose whether the infectious material was located within the nuclei, or whether it became adsorbed to the nuclei during extraction and purification of nuclei.

When crude nuclear fractions were diluted for bioassay in Kühl's (1964) isolation medium (in which nuclei retain their integrity), little infectivty could be demonstrated, whereas when the same homogenates were diluted for bioassay with 0.02 M phosphate buffer, pH 7, infectivity increased conspicuously (Table 27).

When the same experiment was repeated with nuclear fractions that had previously been treated with Triton X-100, the difference in infectivity between inoculum diluted in Kühl's medium and inoculum diluted in phosphate buffer was less pronounced (Table 27).

The low level of infectivity in Kühl's medium was not entirely due to an inhibitory effect of this medium on the infectivity of PSTV. This was shown by the following experiments. Partially purified PSTV (Diener and Raymer, 1969) was suspended in Kühl's medium and diluted for bioassay with the same medium. Another equal amount of PSTV was suspended in, and diluted for bioassay with, phosphate buffer. Still another equal

Table 27. Infectivity of Nuclear Fractions in Kühl's Medium[a] and in Phosphate Buffer

Preparation	Infectivity Indices[b]
Crude nuclear fraction	
Diluted in Kühl's medium	8
Diluted in 0.02 M phosphate buffer, pH 7	71
Triton X-100 treated nuclear fraction	
Diluted in Kühl's medium	17
Diluted in 0.02 M phosphate buffer, pH 7	57
PSTV in Kühl's medium[c]	76
PSTV in 0.02 M phosphate buffer[d]	110
PSTV-added to nuclear fraction[e]	92

From: Diener (1971a).

[a] Kühl (1964).

[b] Preparations assayed undiluted and diluted 10^{-1}, 10^{-2}, 10^{-3}, and 10^{-4}.

[c] Diluted for bioassay with Kühl's medium.

[d] Diluted for bioassay with 0.02 M phosphate buffer, pH 7.

[e] PSTV added to purified nuclei prepared from healthy leaves (suspended in Kühl's medium), diluted for bioassay with Kühl's medium.

portion of PSTV was added to a nuclear fraction (prepared from healthy tissue) and the mixture diluted for bioassay with Kühl's medium. As shown in Table 27, infectivity of PSTV was considerably lower in Kühl's medium than in phosphate buffer, but this difference was not large enough to explain the results with nuclei diluted in the two media (Diener, 1971a).

These results leave little doubt that PSTV is associated with the nuclei (and/or nucleoli) of infected cells. Among the various subcellular fractions, only the nuclear fraction and the tissue debris contained appreciable amounts of infectivity. Kühl (1964) determined that, with his method, approximately 50% of the total number of nuclei appear free in the homogenate. The remaining nuclei presumably are not extracted and may account for the infectivity detected in the tissue debris.

Nuclei suspended in a medium that maintained their integrity expressed little infectivity, whereas nuclei suspended in phosphate buffer were more highly infectious. This observation suggests that the infectious agent may be located within the nuclei, and that it is released from nuclei by the action of phosphate buffer. This conclusion is strengthened by the observation that in successive extractions of nuclei with phosphate buffer, continued but diminishing amounts of infectivity were released (Diener, 1971a).

Further evidence for the intimate association of PSTV with nuclei derived from the observation that infectivity was associated with purified chromatin. As with isolated nuclei, extraction of purified chromatin with phosphate buffer resulted in release of infectivity from the chromatin (Diener, 1971a).

These results were confirmed for CEV by Sänger (1972), who used similar methods for determination of the subcellular location of the viroid and concluded that it was found in nuclei and in association with the chromatin of the host cell.

Semancik *et al.* (1976), in a study of the subcellular distribution of CEV, concluded that the viroid is associated not only with nuclei-rich preparations but also with a plasma membranelike component of the endomembrane system. Evidence of the association of CEV with plasma membrane components was obtained by analyzing 1000 to 80,000 g fractions from CEV-infected *Gynura* on equilibrium sucrose density gradients, followed by bioassay of each collected density gradient fraction for infectivity and for activity of enzyme markers of cell components. Although CEV-infectivity was associated with the plasma membranelike component, the activity of DNA-directed RNA polymerase associated with that component was unusually high. It is known that in higher plants this enzyme is specifically associated with cell nuclei and chloroplasts. Contamination of the plasma membranelike component with nuclear membranes or chloroplast fragments or both could therefore explain the apparent association of infectivity with the plasma component.

The fact that infectious PSTV is located primarily in the nuclei of infected cells does not prove that it is synthesized there. However, experiments with an *in vitro* RNA-synthesizing system, in which purified cell nuclei from infected tomato leaves were used as an enzyme source, demonstrated that this is the case (Takahashi and Diener, 1975; see Section 9.4.2.1). It appears, therefore, that the infecting viroid migrates to the nucleus (by an as yet unknown mechanism) and is replicated there. The absence of significant amounts of PSTV in the cytoplasmic fraction of infected cells suggests that most of the progeny viroid remains in the nucleus.

9.4 MODE(S) OF REPLICATION

In view of the apparent inability of viroids (or their transcription products) to act as messenger RNAs, the biochemical mechanisms by which viroids are replicated, it would seem, are basically different from those operative with conventional viruses. Translation of viral nucleic acid into virus-specified proteins at one stage of the replicative cycle or other is a characteristic feature of all viruses that have been investigated. If, indeed, no viroid-specified proteins are ever synthesized, it follows that viroid replication must be mediated by enzymes already present in host cells before viroid infection. Such enzymes could be either normal host enzymes or they could be synthesized in host cells as a consequence of infection by a conventional plant virus. In the latter case, viroids would be analogous to satellite RNAs that require helper viruses for their own replication. The first question, therefore, is whether viroids are able to replicate autonomously or whether they rely for their replication on products specified by a helper virus.

9.4.1 Autonomous or Helper Virus-Dependent Replication?

If helper viruses are required for viroid replication, all plants that support viroid synthesis must be infected with such entities. Experiments to demonstrate the presence of a helper virus in uninfected Rutgers tomato plants, however, gave negative results (Diener, 1971b). Because every single plant was shown to be susceptible to PSTV, a helper virus, if it exists, must be vertically transmitted through the seed to every single tomato plant. Although seed transmission of certain plant viruses is well known, high rates of transmission are rare and mainly restricted to nematode-borne viruses. Of about 52 viruses known to be seed transmitted, seed transmission approaches 100% only with five viruses (all nematode-borne) and with these only under certain conditions and in certain hosts (Bennett, 1969). With solanaceous plants, such high rates are unknown. Thus no virus is known to be transmitted through the seed of potato and only two viruses are known to be transmitted through the seed of tomato [aside from tomato bunchy top "virus," which is a viroid possibly identical with PSTV (Diener and Raymer, 1971)]. With these two viruses, both of which are nematode-borne, percentage transmission ranges from 1.8% to 19% (Bennett, 1969).

Because PSTV is able to replicate in a number of solanaceous plant species other than potato and tomato (O'Brien and Raymer, 1964), hypothetical helper viruses would have to be present also in these species. Among these are Samsun and Burley tobacco, varieties that for many years have been extensively used in plant virological studies and have been repeatedly examined by electron microscopy of thin sections. No reports of viruslike particles in sections from uninfected tobacco plants have apparently been published. Similarly, electron microscopy of thin sections prepared from healthy and PSTV-infected tomato tissue failed to disclose the presence of virionlike particles in cells of either healthy or infected tissue (R. H. Lawson, personal communication). Sap from uninfected tomato leaves is only weakly antigenic (Bagnall, 1967). Furthermore, double-stranded RNA could be demonstrated in extracts prepared from TMV-infected, but not in extracts from healthy tobacco leaves (Ralph et al., 1965).

None of these observations rule out the possibility that a latent helper virus might be universally present in uninfected tomato and tobacco tissue; they do, however, indicate that such a virus, if it exists, must be present in very small amounts. If so, it appears reasonable to suppose that the amount of helper virus available at infection sites limits PSTV replication, and that addition of helper virus to the inoculum should enhance PSTV infectivity. To determine whether addition of "virus" fractions from tomato leaves to inoculum enhances the infectivity of low molecular weight PSTV, such fractions were isolated from uninoculated tomato leaves by three methods (Diener, 1971b).

Ten-fold dilutions of low molecular weight PSTV (eluted from 10% polyacrylamide gels) were mixed with equal volumes of undiluted "virus" fractions or with 0.02 M phosphate buffer, pH 7, and were bioassayed.

No detectable enhancement of PSTV infectivity occurred, however, when "virus" fractions prepared by any of the three methods used were addded to PSTV prior to inoculation (Diener, 1971b).

Further evidence against the involvement of helper viruses in PSTV replication was obtained by Diener *et al.* (1972). PSTV is able to replicate in a number of solanaceous plant species other than potato and tomato (O'Brien and Raymer, 1964). Thus if PSTV replication depends on the presence of helper viruses, such agents must also have been involved in its replication in these other plant species. Suitable helper entities could already have been present in the test plants prior to inoculation or, since all plants were inoculated with crude extracts from PSTV-infected tissue, they might have beeen introduced into the test plants concurrently with PSTV.

To resolve this ambiguity, Diener *et al.* (1972) repeated the host range studies of O'Brien and Raymer. In addition to inoculation of test plants with crude extracts from PSTV-infected tissue, other test plants were inoculated with PSTV eluted from polyacrylamide gels of small pore size, thus eliminating the possibility of introducing a "conventional" helper virus or its nucleic acid concurrently with PSTV.

The results were unambiguous. All species tested that could be infected by inoculation with crude extracts could also be infected by inoculation with gel eluate, that is, with low molecular weight RNA, thereby strengthening the contention that PSTV is able to replicate in susceptible hosts without the assistance of a conventional helper virus. If such a virus were needed, it would have to occur universally in a wide spectrum of solanaceous plant species, and this helper virus would have to be transmitted through the seed of these species with a much higher frequency than has been reported for a virus in solanaceous plants (Diener *et al.*, 1972).

It appears, therefore, most unlikely that a conventional helper virus is involved in the replication of PSTV, and one is led to conclude that viroid replication is mediated by preexisting normal host enzymes.

A priori, two basic schemes may be envisioned. The first one assumes that viroids are synthesized by the normal RNA synthesizing machinery of the cell; that is, that they are transcribed from DNA templates by host DNA-directed RNA polymerases. The second scheme assumes that viroid replication is DNA independent and that it occurs from RNA templates synthesized by host enzymes that are capable of functioning as RNA-directed RNA polymerases. In this case, the mode of viroid replication would be analogous to that of many RNA viruses, except that this replication would be mediated entirely by normal host enzymes.

On the basis of available evidence, no conclusive distinction between these two theoretically possible mechanisms is possible. Consequently, data that appear to favor one or the other mode of viroid replication are presented, but no definitive conclusion is stated.

9.4.2 Evidence Favoring Deoxyribonucleic Acid-Directed Replication

If viroids are transcribed from DNA templates, two schemes may be envisioned. DNA sequences complementary to viroids could either be present in repressed form in uninfected host plants, or such sequences could be produced as a consequence of viroid infection. In the former case, the infecting viroid would have to act, directly or indirectly, as a regulatory molecule by derepressing viroid-specifying sequences in the host genome. In the latter case, RNA-directed DNA polymerases (reverse transcriptases) would have to exist as normal host enzymes in all plant species in which viroids are capable of replication. Two lines of evidence appear to favor the concept that viroid replication is DNA-directed.

9.4.2.1 Sensitivity to Actinomycin D

Actinomycin D is well known to specifically inhibit RNA transcription from DNA templates in animal cells and not to seriously interfere with the replication of many animal virus RNAs. A number of reports indicate that actinomycin D similarly inhibits RNA transcription in higher plants and does not seriously interfere with the replication of several plant viral RNAs (Sänger and Knight, 1963; Bancroft and Key, 1964; Semal, 1966; Babos and Shearer, 1969; Romero, 1972; Reunova et al., 1973). At early stages of infection, however, actinomycin D has been shown to inhibit the replication of certain plant viruses (Bancroft and Key, 1964; Lockhart and Semancik, 1968, 1969) and to stimulate replication of at least one plant virus (Reunova et al., 1973).

To determine whether PSTV replication is or is not sensitive to actinomycin D, Diener and Smith (1975) vacuum-infilrated water or solutions of actinomycin D into leaf strips from healthy or PSTV-infected tomato plants, then incubated the leaf strips in solutions of [³H]uracil, extracted total nucleic acids, and studied [³H]uracil incorporation into host RNA and PSTV.

Figure 40*A* shows that with healthy leaves, in the absence of pretreatment with actinomycin D, three major radioactive peaks were evident. These peaks coincided with the positions in the gels of tRNA and of the two cytoplasmic ribosomal RNAs. Figure 40*B* reveals that pretreatment of leaf strips with actinomycin D resulted in almost complete inhibition of [³H]uracil incorporation, except in the region of tRNA, where low levels of incorporation were consistently observed after actinomycin D pretreatment. Figure 41 illustrates an analogous experiment made with leaf strips from PSTV-infected plants. Again [³H]uracil incorporation occurred into tRNA and the two cytoplasmic ribosomal RNA compounds in the absence of actinomycin D pretreatment (Fig. 41*A*); pretreatment with the antibiotic largely inhibited this incorporation (Fig. 41 *B*).

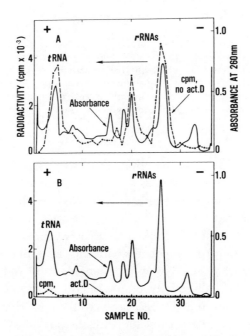

Figure 40.   Ultraviolet-absorption (———) and radioactivity (•······•) profiles of nucleic acid preparations from healthy tomato leaves after electrophoresis in 2.4% polyacrylamide gels for 3 hours at 4°C (5 mA/tube, constant current). Electrophoretic movement is from right to left. (*A*) leaf strips infiltrated with water, followed by incubation for 8 hours at 25°C with 125 μCi of [³H]uracil; (*B*) leaf strips infiltrated with 30 μg/ml of actinomycin D, followed by incubation as in A. From: Diener and Smith (1975).

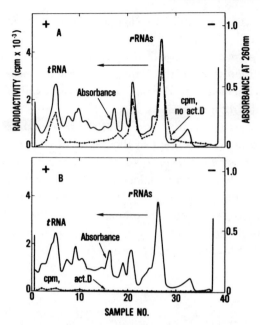

Figure 41. Ultraviolet-absorption (———) and radioactivity (•·····•) profiles of nucleic acid preparations from PSTV-infected tomato leaves after electrophoresis in 2.4% polyacrylamide gels. (*A*) leaf strips infiltrated with water, followed by incubation for 8 hours at 25°C with 125 μCi of [³H]uracil; (*B*) leaf strips infiltrated with 30 μg/ml of actinomycin D, followed by incubation as in (*A*). From: Diener and Smith (1975).

To detect [³H]uracil incorporation into PSTV, low molecular weight RNA preparations were analyzed, together with added unlabeled PSTV, in high percentage polyacrylamide gels. Results are shown in Figure 42 (Diener and Smith, 1975).

The radioactivity profile of low molecular weight RNAs from healthy leaves (Fig. 42*A*) showed components with the electrophoretic mobility of 5 S RNA and of at least three minor components of cellular RNA, but not of PSTV. The radioactivity profile of low molecular weight RNAs from PSTV-infected leaves (Fig. 42*B*) showed the same components but, in addition, a component that coincides with the position of marker PSTV. This coincidence and the absence of the component in identically treated preparations from healthy leaves was considered strong evidence that this component was the result of [³H]uracil incorporation into PSTV (Diener and Smith, 1975).

Pretreatment of leaf strips from healthy plants with actinomycin D

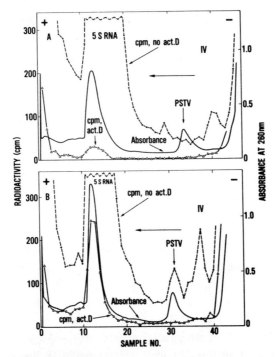

Figure 42.   Ultraviolet-absorption (————) and radioactivity profiles of low molecular weight RNA preparations from healthy or PSTV-infected tomato leaves after electrophoresis in 20% polyacrylamide gels for 7.5 hours at 4°C. Electrophoretic movement is from right to left. (*A*) leaf strips from healthy plants infiltrated with water (•······•) or with 30 μg/ml of actinomycin D (△·····△), followed by incubation for 8 hours at 25°C with 125 μCi of [³H]uracil. Unlabeled purified PSTV was added as internal marker; (*B*) leaf strips from PSTV-infected leaves treated as described under A; IV = unidentified minor component of cellular RNA (see Diener, 1972a). From: Diener and Smith (1975).

resulted in almost complete inhibition of subsequent [³H]uracil incorporation into low molecular weight RNAs (Fig. 42A). Pretreatment of leaf strips from PSTV-infected plants with actinomycin D resulted in a similar reduction of [³H]uracil incorporation into low molecular weight RNAs (incorporation into 5 S RNA, although considerable, was only 5% of that into 5 S RNA in leaf strips not pretreated with actinomycin D) (Fig. 42B). No radioactive peak is discernible in the PSTV region of the gel and calculation indicated that incorporation into PSTV could not have been larger than 15% of that into PSTV from leaves not pretreated with actinomycin D (Diener and Smith, 1975).

These results clearly indicate that the replication of PSTV is sensitive to actinomycin D and the known specificity of the antibiotic suggests that the viroid may be transcribed from a DNA template.

Alternatively, however, it is possible that the inhibition of PSTV replication is due to a nonspecific effect of actinomycin D on cellular metabolism, resulting in depletion of a precursor that is essential for RNA synthesis. For example, continued synthesis of a cellular RNA might be required for PSTV replication (Diener and Smith, 1975). Specifically, such a required host RNA or oligonucleotide might serve as a primer for PSTV replication.

Actinomycin D sensitivity of PSTV replication has been demonstrated also in a cell-free RNA synthesizing system. Takahashi and Diener (1975) isolated and purified nuclei from healthy and PSTV-infected tomato leaves and used them, together with the necessary RNA precursors and other constituents, to assemble an *in vitro* RNA synthesizing system. Table 28 summarizes several properties of this system, and Table 29 characterizes the reaction product as RNA (Takahashi and Diener, 1975). Pretreatment with actinomycin D inhibited RNA synthesis. Isolation of low molecular weight RNAs from the *in vitro* reaction mixtures and analysis by gel electrophoresis revealed, in the absence of actinomycin D, incorporation into a component with the electrophoretic mobility of PSTV in reactions per-

Table 28.   Requirements for Ribonucleotide Polymerization by Nuclei

System	cpm/Assay	
	Uninfected	Infected
Complete[a]	11,141	12,016
minus ATP	1,896	1,406
minus CTP	1,938	1,798
minus GTP	1,586	1,064
minus ATP, CTP, GTP	960	920
minus PEP, pyruvate kinase	1,574	1,062
Complete	10,127	10,070
minus MgCl$_2$	385	330
minus 2-mercaptoethanol	3,901	3,726
minus 2-mercaptoethanol, plus 4 mM dithiothreitol	9,246	10,638

From: Takahashi and Diener (1975).
[a] Complete reaction mixture for standard assay. Each assay contained 5 x 10^5 nuclei. Incubation: 60 minutes, 30°C.

Table 29. **Characteristics of the Reaction Product of Ribonucleotide Polymerization by Nuclei**

	cpm/Assay	
Condition[a]	Uninfected	Infected
In vitro synthesized ³H-labeled nuclear RNA	10,242	9,693
Treated with 1 *N* NaOH at 37°C for 18 hours, neutralized, and acidified with 5% TCA	253	242
Treated with 5% TCA at 70°C for 20 minutes	8,283	8,265
Incubated with RNase (50 μg in 5 mM MgCl₂) for 60 minutes	868	860
Incubated with DNase (50 μg in 5 mM MgCl₂) for 60 minutes	9,960	9,633

From: Takahashi and Diener (1975).

[a] After incubation for 60 minutes at 30°C, 500 μg of soluble yeast RNA was added as carrier to each reaction mixture. TCA-insoluble residues were treated as described.

formed with nuclei from PSTV-infected leaves, but not in reactions with nuclei from healthy leaves (Fig. 43).

Preincubation of nuclei with actinomycin D resulted in complete inhibition of incorporation into PSTV (Fig. 43) (Takahashi and Diener, 1975). Thus in confirmation of the results of *in vivo* studies (Diener and Smith, 1975), PSTV replication in this *in vitro* system is sensitive to treatment with actinomycin D.

9.4.2.2 Viroid Hybridization to Host Deoxyribonucleic Acid

The detection in molecular hybridization experiments of DNA sequences complementary to viroids has been claimed in two reports.

Semancik and Geelen (1975) hybridized *in vitro* ¹²⁵I-labeled CEV to nucleic acids isolated from various subcellular fractions derived from healthy or CEV-infected *Gynura aurantiaca* plants. The most significant level of RNase-resistant ¹²⁵I-CEV binding was observed with nucleic acid from a 250-*g* pellet which contained the bulk of the nuclear DNA as judged by a colorimetric test. To determine whether a CEV-like DNA is specific to CEV-infected tissue, high molecular weight DNAs were prepared from healthy and CEV-infected *Gynura* and tomato, as well as from CEV-resistant tobacco (*Nicotiana tabacum*), and cowpea (*Vigna unguicul-*

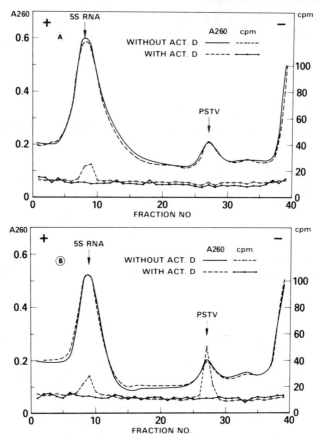

Figure 43. Ultraviolet-absorption and radioactivity distribution of low molecular weight RNAs isolated from incubation mixtures, each containing 8 μCi of [3H]UTP and nuclei equivalent to 20 μg of DNA, pretreated with either water or 50 μg/ml of actinomycin D, after electrophoresis in 20% polyacrylamide gels for 7.5 hours at 4°C. Migration of RNAs is from right to left. DNA and high molecular weight RNAs were removed prior to electrophoresis. Conditions of electrophoresis were such that transfer RNA was run off the gels. Incorporation of [3H]UTP into 5 S RNA and PSTV amounted to 0.05 to 0.2% of total incorporation into RNA. (*A*) Nuclei isolated from uninfected leaves; (*B*) nuclei isolated from PSTV-infected leaves. From: Takahashi and Diener (1975).

ata cv. Early Ramshorn). The results indicated that 125I-CEV specifically hybridizes only with DNA extracted from diseased tissue (Semancik and Geelen, 1975). The authors concluded that the CEV-specific DNA located in the 250-*g* pellet supports a nuclear phase in the synthesis of the viroid. They stated that the question as to whether the DNA complementary to CEV is synthesized *de novo* by the action of an induced RNA-directed

DNA polymerase, or exists in a limited number of copies and is amplified by normal DNA polymerase activity in the process of pathogenesis, remains to be determined.

Unfortunately, the significance of these findings is uncertain because (1) hybridization reactions were made with "DNA-rich" preparations, and not with purified DNA, (2) the percentage of input CEV hybridized was small (3.5% and 2.9%, for DNA-rich preparations from CEV-infected tomato and *Gynura*, respectively), and (3) neither C_0t values for hybrid formation nor thermal denaturation properties of the hybrids were reported. Evidently, a relatively minor contamination of the purified and iodinated CEV preparations with cellular RNA could explain the reported results.

Semancik and Geelen's (1975) results are at variance with results obtained by hybridization of ^{125}I-labeled PSTV to host DNA (Hadidi *et al.*, 1976), which showed that infrequent, if not unique, DNA sequences complementary to PSTV appear to exist in both uninfected and infected cells of PSTV host plants.

DNA titration experiments (Fig. 44) indicated that at least 60% of PSTV is represented by complementary sequences in the DNAs of several

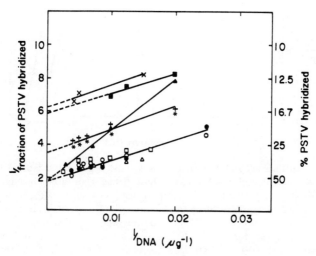

Figure 44. Double reciprocal plot from hybridization of a constant amount of ^{125}I-labeled PSTV with an increasing amount of cellular DNA of phylogenetically diverse plants at a C_0t value of 2×10^4. Sources of DNA: o-o-o, tomato; •-••-•, PSTV-infected tomato; □-□-□-□, potato; △-△-△, *Physalis*; ▲-▲-▲, bean; +-+-+, *Gynura*; *-*-*, PSTV-infected *Gynura*; ■-■-■, Chinese cabbage; and x-x-x, barley. From: Hadidi *et al.* (1976).

uninfected solanaceous host species. The DNA of a nonsolanaceous host, *Gynura aurantiaca,* appeared to contain DNA sequences related to only about 30% of the PSTV molecule, and the DNAs of Chinese cabbage and barley, which are nonhosts of PSTV, appear to contain sequences to only a small portion of PSTV or no such sequences (Hadidi *et al.,* 1976). Surprisingly, however, the DNA of Black Valentine bean, which also is resistant to infection with PSTV, appeared to contain DNA sequences complementary to a large portion of the PSTV molecule (Fig. 44). In general, the more distant phylogenetically plant species are from solanaceous plants, the fewer PSTV-related sequences their DNAs appear to contain. Also, study of the thermal denaturation properties of the hybrids revealed that those consisting of PSTV and DNA from solanaceous host plant melted over a relatively narrow range of temperatures (Fig. 45), giving no evidence of appreciable base mismatching, whereas hybrids formed of

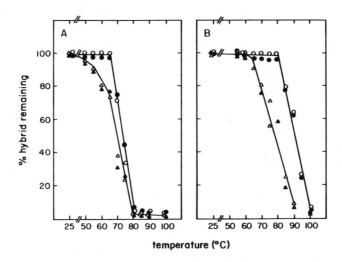

Figure 45. (*A*) Thermal stability of hybrids formed between ^{125}I-labeled PSTV and DNA from uninfected and PSTV-infected host tissue. Hybridization mixtures, each containing 500 μg of DNA and 5 ng of ^{125}I-labeled PSTV in 0.4 *M* sodium phosphate buffer, were boiled in capillary tubes for 5 minutes, and were then incubated at 60°C to a C_0t value of 1 × 10⁴. Hybridization mixtures were expelled into 0.1 × SSC and aliquots were exposed for 5 minuts to increasing increments of temperature. o-o-o, tomato DNA; •-•-•, PSTV-infected tomato DNA; △-△-△, *Gynura* DNA; ▲-▲-▲, PSTV-infected *Gynura* DNA. (*B*) Thermal stability of hybrids formed between ^{125}I-labeled PSTV and DNA or RNA of uninfected or PSTV-infected tomato measured in 1 × SSC. t_m was measured as described in (*A*), except that hybridization mixtures were expelled into 1 × SSC. o-o-o, tomato DNA; •-•-•, PSTV-infected tomato DNA; △-△-△, tomato RNA; ▲-▲-▲, PSTV-infected tomato RNA. From: Hadidi *et al.* (1976).

PSTV and DNA from *Gynura* melted over a wider temperature range (Fig. 45), giving evidence of considerable base mismatching. As shown in Fig. 44, the hybridization kinetics between PSTV and DNAs from solanaceous host plants were identical, irrespective of whether DNA from healthy or PSTV-infected plants was used. Thus no evidence for the formation of new DNA sequences complementary to PSTV was obtained. Consequently, if progeny PSTV is indeed transcribed from these DNA sequences, the incoming viroid would have to act as a derepressor of the PSTV sequences.

Although the high t_m value of hybrids formed between PSTV and DNA from solanaceous host plants is compatible with the G + C content of PSTV (about 55%, see Section 8.3.1), the possibility had to be considered that the hybrids found were not complexes between plant DNA and PSTV, but between DNA and cellular RNA contaminants which may have been present in the purified PSTV preparations and which had been labeled with ^{125}I concomitantly with PSTV. This appeared unlikely, because fingerprinting of ^{125}I-labeled PSTV showed the preparations to be at least 90% pure (Dickson *et al.*, 1975), whereas as much as 35% of the input RNA hybridized to DNA (Hadidi *et al.*, 1976). No competition experiments with unlabeled PSTV or cellular RNA were made, however.

9.4.2.3 *Deoxyribonucleic Acid-Directed Ribonucleic Acid Polymerases in Healthy and Viroid-Infected Plants*

Geelen *et al.* (1976) investigated properties of RNA polymerases in healthy and CEV-infected *Gynura* plants. In reactions similar to those performed by Takahashi and Diener (1975) in which isolated nuclei from healthy and viroid-infected plants were used, the authors showed that the product was RNase-sensitive and that the addition of exogenous substrate in the form of purified CEV, single-stranded RNA of bean pod mottle virus, or double-stranded RNA of bacteriophage $\phi 6$ to intact nuclei pretreated with DNase, did not affect the level of [³H]UMP incorporation. RNA polymerases solubilized from nuclear preparations were separated by chromatography into three fractions, two of which preferred native DNA as template, whereas a third preferred denatured DNA. All three enzyme fractions lacked specificity, because all the RNAs tested, including CEV, were accepted as substrates (Geelen *et al.*, 1976). The products of the reaction with CEV and either one of two of the fractions as enzyme source were shown to be of low molecular weight (5 S and smaller). Because the solubilized RNA polymerases, isolated from both healthy and CEV-infected *Gynura* plants, accepted CEV as well as or better than native DNA, Geelen *et al.* (1976) concluded that probably no *de novo* synthesis of polymerase is required for CEV synthesis and that *in vivo* synthesis of viroid is accomplished via a DNA template.

9.4.2.4   Speculative Model of Viroid Replication

The finding of DNA sequences complementary to PSTV and CEV in viroid host plants, as well as the demonstrated sensitivity of PSTV replication to actinomycin D, led Diener (1977) to propose a speculative model of viroid replication. This model postulates that viroid replication is DNA directed; that is, that viroids are transcribed from normally repressed host genes that are activated as a result of viroid infection. The author recognized that although evidence available at the time seemed to favor this concept, this evidence was far from conclusive, and that therefore future work may invalidate the model or necessitate major modification.

The model is based on the familiar scheme proposed by Britten and Davidson (1969) for gene regulation in higher cells, in which producer genes are derepressed by activator RNAs which form sequence-specific complexes with receptor genes linked to producer genes. Activator RNAs are transcribed from integrator genes which, in turn, are regulated by sensor genes (Britten and Davidson, 1969).

Comparison of the properties of viroids with those postulated for activator RNAs show that viroids immediately fulfill three of the four criteria: (1) they are, in the main, confined to the nucleus, (2) they are associated with chromatin, and (3) they are not detectable in ribosomes or the cytoplasm of infected cells.

The fourth criterion for activator RNAs is that they are often the product of the redundant fraction of the genome (Britten and Davidson, 1969). Although DNA-viroid hybridization studies appear to show that this is not true for viroids (Hadidi *et al.*, 1976), closer examination reveals that viroids do fulfill this criterion. Native viroids exist in the form of hairpinlike structures with extensive regions of intramolecular base-pairing (see Chapter 7). It follows, as already pointed out (Reanney, 1976), that viroids are most likely transcribed from palindromic regions of DNA; indeed, the physical properties of viroids are to a large extent those expected of the transcriptional product of an imperfect palindrome. Palindromic structures are known to exist in the genomes of higher plants (Walbot and Dure, 1976).

Because of the rapidity with which intramolecular complementary DNA sequences reanneal, palindromic regions anneal in DNA reassociation experiments as if they were highly reiterated (Walbot and Dure, 1976), yet each region may contain a unique or infrequent sequence. Thus there is no inherent conflict between the observation that DNA sequences complementary to PSTV are infrequent or unique ones (Hadidi et al., 1976) and the postulate that viroids are activator RNAs in Britten and Davidson's sense, with the implication that they are the product of the redundant fraction of the genome.

It is not known whether the mechanism of gene derepression involves, as has often been suggested, the dissociation of nuclear protein from DNA, but, as noted (McClements and Kaesberg, 1977), the hairpinlike structure of viroids provides the proper configuration for specific interaction of the RNA with a protein (Gralla et al., 1974).

On theoretical grounds, Gierer (1973) has postulated that "there are many regulatory proteins, each capable of activating its own cistron, because each cistron is linked to the recognition sequence for the corresponding regulatory protein. In this way, a stage of regulation would be self-stabilizing, via the cytoplasm, and could be transferred to daughter cells." The same purpose could be achieved in a more direct fashion, however, if gene regulation were mediated by RNA instead of protein, that is, if activator RNAs could induce their own synthesis, because the whole process could take place in the nucleus without the necessity for migration of messenger RNAs from the nucleus to the cytoplasm and of the newly synthesized regulatory proteins back into the nucleus. Diener (1977) postulated that viroids represent such autoinducing regulatory RNAs.

As a corollary of this concept, one would expect that viroids propagated in different host species would have somewhat different primary structures. As discussed in Section 9.4.3.3, however, more recent evidence indicates that this is not the case and that, to the contrary, the incoming viroid and not the host supplies the information for the primary sequence of progeny viroid. Also, the finding of RNA sequences complementary to CEV (see Section 9.4.3.2) weakens the case for DNA-directed viroid synthesis and makes it appear more plausible that viroid replication may be RNA-directed. Viroids, nevertheless, may act as activator RNAs in Britten and Davidson's (1969) sense and, by their interference with gene regulation, bring about the aberrations of host metabolism that, in certain hosts, result in symptom formation (see Section 9.5).

9.4.3 Evidence Favoring Ribonucleic Acid-Directed Replication

In contrast to the postulate of DNA-directed viroid replication, which necessitates either novel and unorthodox assumptions (see Section 9.4.2.4) or the presence, in uninfected host plants, of reverse transcriptaselike enzymes [which have not so far been identified in cells of higher plants (Diener and Hadidi, 1977)], the postulate of RNA-directed viroid replication does not require unorthodox assumptions, provided that enzymes capable of transcribing RNA from RNA templates occur in uninfected cells of host plants. The identification of such enzymes is the topic of the next section.

9.4.3.1 *Ribonucleic Acid-Directed Ribonucleic Acid Polymerases in Higher Plants*

Astier-Manifacier and Cornuet (1971) were apparently the first to isolate an RNA-directed RNA polymerase from a higher plant source not obviously infected by a virus. The authors purified the enzyme from the chloroplast fraction of Chinese cabbage leaves and showed that the enzyme incorporates nucleotides from ribonucleoside triphosphates into an acid-insoluble polymer, that addition of RNA template is required for enzyme activity, and that turnip yellow mosaic virus RNA is among the most active templates. The product is 85% RNase-resistant and the label cannot be displaced by viral RNA. The authors reported that the enzyme is found in apparently healthy plants and that its specific activity increases during the first week of infection by turnip yellow mosaic virus. They concluded that the synthesized RNA is complementary to the template (Astier-Manifacier and Cornuet, 1971).

Later, Duda *et al.* (1973) reported on an enzyme with similar properties in the 30,900-*g* supernatant fraction of a homogenate prepared from apparently healthy tobacco leaves. As in the case of the enzyme from Chinese cabbage, the tobacco enzyme requires the presence of a divalent metal ion and four ribonucleotide triphosphates for activity. The reaction product is double-stranded RNA as evidenced by its resistance to RNase and sharp thermal transition; it is of low molecular weight as determined by gel electrophoresis (Duda *et al.*, 1973). Addition of various RNA species stimulates the reaction but, in contrast to the Chinese cabbage enzyme, the added RNA does not appear to be associated with the enzyme product or to act either as template or primer. Duda *et al.* (1973) further reported that the enzyme can be isolated from apparently healthy plants, but that it is also found in tobacco mosaic virus-infected plants and that its specific activity increases after infection.

The authors noted that the occurrence of an RNA-directed RNA polymerase in normal cells is implicit in a scheme proposed by Diener (1972b) for the replication of viroids and that the newly discovered enzyme might be an obvious candidate. No evidence exists, however, for the role of this enzyme in viroid replication.

Further evidence for the presence of RNA-directed RNA polymerases in uninfected plants has been presented by several workers (Fraenkel-Conrat, 1976; Bol *et al.*, 1976; Stussi-Garaud *et al.*, 1977; Le Roy *et al.*, 1977; White and Murakishi, 1977; Ikegami and Fraenkel-Conrat, 1978; Romaine and Zaitlin, 1978). It now appears certain that such enzymes are widely distributed, if not universally present, in uninfected higher plants. Although enzymes with specificities similar to those isolated from plants have been reported to occur in certain apparently uninfected animal cells (Downey *et al.*, 1973; Mikoshiba *et al.*, 1974), these reports have not been confirmed or further substantiated. It is possible, therefore, that such enzymes are restricted to higher plants (Ikegami and Fraenkel-Conrat, 1978) and that this might be the reason why viroids have not been found in animal cells (see Chapter 10). The involvement of host-specific, and preexisting, RNA-directed RNA polymerases in viroid replication, however, is not established and, because of two properties of these enzymes, such involvement appears, in fact, unlikely. In the first place, some data indicate that viroid replication occurs in the nucleus of infected cells (see Section 9.4.2.1), yet the nuclear fraction (1000 g pellet) from healthy tobacco leaves is essentially devoid of RNA-directed RNA polymerase activity (Le Roy *et al.*, 1977; Ikegami and Fraenkel-Conrat, 1978). In the second place, the host enzyme(s), as isolated from healthy plants, appears to be capable of synthesizing only RNA complementary to the template RNA; that is minus strand RNA (Romaine and Zaitlin, 1978); no progeny plus strands have been detected. However, neither of these observations conclusively rules out a function of host RNA-directed RNA polymerase in viroid replication. Contrary to present indications, viroids might be replicated in the cytoplasm, for example. Also, it is conceivable that during extraction and partial purification, a portion of the enzyme essential for the synthesis of plus strands, but not for the completion of minus strands, is lost from the preparation.

An enzyme with properties similar to the ones prepared from Chinese cabbage and tobacco was detected in healthy tomato leaves; enzyme activity is greatly enhanced by the addition of PSTV, as well as by several other RNA species, but not by addition of DNA (Hadidi, unpublished; quoted in Diener and Hadidi, 1977). Again, the role of this enzyme, if any, in PSTV replication is unknown.

9.4.3.2   Ribonucleic Acid Complementary to Viroid

In efforts to detect RNA sequences complementary to PSTV, Hadidi *et al.* (1976) hybridized ^{125}I-labeled PSTV to total RNA from PSTV-infected or uninfected tomato leaves. In all tests, the hybrid yields were small and did not significantly increase when the ratio of RNA to a constant amount of ^{125}I-PSTV was increased. The percentage of hybridization in complexes formed between labeled PSTV and RNA from PSTV-infected leaves was consistently higher than that obtained with complexes formed between labeled PSTV and RNA from healthy leaves. This difference was not large enough, however, to account for RNA sequences complementary to a large portion of PSTV. Furthermore, all RNA•RNA complexes formed melted over a broad range of temperature (Fig. 45B), indicating a large degree of mismatched duplex formation (Hadidi *et. al.*, 1976). The authors' failure to detect RNA sequences complementary to a major portion of PSTV may have been the result of competition between labeled PSTV and a great excess of unlabeled PSTV in the cellular RNA preparations.

Grill and Semancik (1978), on the other hand, obtained evidence for the existence, in CEV-infected *Gynura* leaves, of RNA sequences complementary to this viroid. Molecular hybridization between ^{125}I-labeled CEV and various nucleic acid fractions from healthy and CEV-infected *Gynura* indicated that significant RNase-resistant labeled CEV was detectable only in hybridization experiments with nucleic acid fractions from infected tissue. The largest amount of hybridization occurred with the LiCl-precipitated nucleic acids of the 100,000 × g supernatant material, suggesting that the labeled CEV hybridizes to a single-stranded RNA. Significant amounts of hybrids were formed with both the LiCl precipitate and supernatant material of the 750 × g nuclei-rich fraction. Because of the presence of nucleic acid species hybridizable to CEV in different subcellular fractions, the authors concluded that possibly nucleic acids with different properties but complementary to CEV may exist. The LiCl-precipitable complementary nucleic acid must be distinct from CEV, because the latter remains predominantly in the LiCl supernatant fraction (Grill and Semancik, 1978). Characterization of the hybridization reaction demonstrated that the interaction between CEV and nucleic acids from infected host tissue was probably specific and that the hybrid formed had the expected properties of an RNA·RNA duplex, as evidenced by its resistance to RNase H and its t_m value of 90°C in 0.01 × SSC (Grill and Semancik, 1978).

Although the role of the complementary RNA is unknown, the authors suggested that it might be involved in the replication of CEV or that it might have a regulatory role and be involved in pathogenesis, or both. The authors did not, however, clarify the relation, if any, between the DNA sequences complementary to CEV that were reported to exist earlier (Semancik and Geelen, 1975) and the complementary RNA sequences described in the later report (Grill and Semancik, 1978).

9.4.3.3 Progeny Viroid is Copy of Infecting Viroid

As discussed in Section 9.4.2.4, if viroids were transcribed from preexisting DNA templates, one might expect that viroids propagated in different hosts would have somewhat different primary structures, and it would be formally possible that the viroid genome could undergo major alterations upon replication in different hosts. If, on the other hand, viroid replication were DNA-independent, that is, if viroids were transcribed from complementary RNA strands produced after infection with the incoming viroid serving as template, such major alterations in the primary structure evidently could not occur. Thus determination, in different hosts, of the primary structure of progeny viroids should give clues as to the mode of viroid replication.

Two different approaches to this problem have yielded results indicating that the primary structure of viroids is more or less faithfully maintained, irrespective of the host in which the pathogen is propagated.

The first approach was taken in a study by Dickson et al. (1978), in which two viroids, PSTV and CEV, were each propagated in two different hosts, tomato and Gynura, the progeny viroids purified, labeled with ^{125}I, and their primary structures studied by RNA fingerprinting of RNase T_1 digests. Because earlier studies had shown a clear distinction of the fingerprint patterns of PSTV propagated in tomato and CEV propagated in *Gynura* (Dickson *et al.*, 1975), it was of interest to determine whether the fingerprint patterns of PSTV propagated in *Gynura* resembled that of CEV from *Gynura* or whether it resembled that of the infecting viroid, tomato PSTV, and conversely, whether the fingerprint of CEV propagated in tomato resembled that of PSTV from tomato or whether it resembled that of *Gynura* CEV.

The results of the experiments (Dickson *et al.*, 1978) clearly indicated that PSTV did not undergo major sequence changes when propagated in *Gynura* and, similarly, that CEV displayed essentially the same fingerprint pattern, irrespective of whether the viroid was propagated in *Gynura* or tomato. Minor variations in the viroid fingerprints were observed but their source has not so far been elucidated (Dickson *et al.*, 1978). It is thus still possible that the host exerts a minor influence on the primary structure of viroids, but, alternatively, these minor variations may also be due to other causes.

The results are compatible with those of experiments that demonstrate that PSTV from tomato is able to infect chrysanthemum plants and to produce a disease indistinguishable from that caused by CSV, but yielding a progeny viroid with an electrophoretic mobility in gels identical with that of PSTV from tomato rather than that of CSV [Diener and Smith, unpublished; cited in Diener and Hadidi (1977)]. Again, it appears that the properties of progeny viroids are largely determined by the infecting viroid and not by the particular host in which the viroid is propagated.

The second approach uses complementary DNA as a probe in molecular hybridization experiments. Owens (1978) demonstrated that DNA complementary to PSTV (PSTV·cDNA) is produced in an *in vitro* reaction with avian myeloblastosis virus reverse transcriptase and hydrolyzed calf thymus DNA as primer. After purification, the PSTV·cDNA was shown to hybridize specifically with PSTV or low molecular weight RNA preparations from PSTV-infected plants, but not with cellular RNAs or with low molecular weight RNA preparations from healthy plants (Owens, 1978).

Hybridization reactions of PSTV·cDNA with PSTV propagated in tomato, tobacco, *Gynura*, or chrysanthemum did not disclose any differences in the extent of hybridization, thus indicating that the viroid does not undergo any major sequence alterations if propagated in any one of these hosts (Owens *et al.*, 1978).

These experiments clearly demonstrate that the nucleotide sequence of progeny viroids is largely determined by that of the infecting viroid and that the host has no influence or at most a minor one on the primary structure of the progeny viroid.

9.4.3.4   A Model of Ribonucleic Acid-Directed Viroid Replication

All studies discussed in this section seem to favor—but do not prove—a model that assumes DNA independent, that is, RNA-directed viroid replication. Enzymes capable of such RNA synthesis are present in host plants of viroids, RNA sequences complementary to at least one viroid have been identified, and the nucleotide sequences of progeny viroids have been shown to closely resemble those of the infecting viroid, irrespective of the particular host in which the viroid is propagated. If so, why is viroid replication sensitive to actinomycin D? As discussed in Section 9.4.2.1, continued synthesis of a cellular RNA may be required for viroid replication to occur. Most plausibly, such a host RNA or oligonucleotide may be required to serve as primer for viroid replication.

9.4.4   Replication of Circular and Linear Viroid Forms

The *in vivo* relationship of circular to linear viroid forms has been studied in experiments in which the *de novo* formation of either was determined by *in vivo* ^{32}P labeling of viroid.

Sänger and Ramm (1975) showed that, under suitable conditions, *in vivo* labeling of PSTV and CEV is possible but concluded that the "long term" labeling experiments "for systemically infected whole plants and individual leaves do not allow any statements on the actual mode of viroid replication for which actual pulse-labeling is needed" (Sänger and Ramm, 1975).

Hadidi and Diener (1977), on the basis of similar *in vivo* ^{32}P incorporation experiments with PSTV, concluded that *de novo* synthesis of the viroid can be studied *in vivo* by the introduction of ^{32}P through the roots of PSTV-infected tomato plants, followed by isolation and fractionation of RNA and gel electrophoretic analysis of labeled low molecular weight RNA. Optimal labeling of PSTV occurred when the plants were allowed to take up ^{32}P at the time when systemic symptoms started to appear.

These techniques were used to study the *in vivo* formation of circular and linear PSTV molecules (Hadidi and Diener, 1978). As discussed in Chapter 7, both circular and linear forms of PSTV have been detected by electron microscopy of purified denatured viroid preparations. For reasons already discussed, it appears unlikely that the linear and circular molecules are distinct RNA species; it appears more plausible that they represent two stages of maturity of a single RNA species. The results of Owens *et al.* (1977), who were able to separate linear from circular RNA and to show that both forms are infectious, strongly support this contention. Also, Hadidi and Diener (1978) showed that ^{32}P-labeled circular and linear PSTV molecules have identical nucleotide compositions and Hadidi and Keith (1978) showed that fingerprint patterns of RNase T_1-resistant oligonucleotides of uniformly ^{32}P-labeled circular and linear PSTV are very similar.

The possibility, nevertheless, remained that linear PSTV molecules might arise *in vitro* by breakage of circular PSTV molecules during viroid extraction and purification. Such an explanation has been advanced by Sänger *et al.* (1976) to explain the presence of linear molecules in purified viroid preparations.

To determine whether linear PSTV is formed *in vivo* and, if so, whether it is synthesized independently of circular PSTV or whether one arises from the other, Hadidi and Diener (1978) allowed PSTV-infected plants to take up ^{32}P and to incorporate it for various periods of time beginning on the first day of symptom appearance. Periods of ^{32}P incorporation as short as 2 hours were sufficient to detect newly formed circular and linear PSTV. As shown in Table 30, with 2 hours of incorporation, labeling of circular PSTV prevailed but, with increasing duration of ^{32}P incorporation, the proportion of labeled circular to linear PSTV steadily decreased until, with 10 days of incorporation, labeled linear PSTV prevailed (Hadidi and Diener, 1978).

These results seem to rule out breakage of circular PSTV during extraction and purification of RNA as a major source of linear PSTV, because all extractions were made in an identical fashion yet widely varying proportions of circular to linear RNA were obtained.

The results rather suggest that linear PSTV may be formed *in vivo* by specific cleavage of circular PSTV. Other explanations, however, are possible. For example, the prevalance of labeled linear PSTV after long periods of ^{32}P incorporation might be a consequence of radiation-induced breakage of circular PSTV. This possibility was ruled out by the demonstration that similar results were obtained when ^{32}P was replaced by [5,6-^{3}H]uridine, or by tritiated adenosine, cytidine, or guanosine (Hadidi and Diener, 1978).

Circular and linear PSTV might be synthesized independently, with cir-

Table 30. Relationship between duration of [32]P Incorporation and the Relative Proportions of C-PSTV and L-PSTV[a]

Time Allowed for Incorporation of [32]P (hours)	Circular-PSTV (%)	Linear-PSTV (%)
2	71	29
4	67	33
6	63	37
8	56	44
12	54	46
24 .	52	48
48	49	51
96	46	54
120	44	56
168	40	60
240	≤ 30	≥ 70

From: Hadidi and Diener (1978).

[a] At the first day of symptom expression, groups of six PSTV-infected tomato plants were allowed to incorporate 5 mCi of ^{32}P per group for periods of time ranging from 2 to 240 hours. Isolated ^{32}P-labeled PSTV was analyzed in 5% polyacrylamide gels under denaturing conditions and total counts of radioactive materials in the region of C- or L-PSTV were determined. After subtracting the counter background, the percentage of counts in each was calculated.

cular PSTV synthesis dominating at the time of first symptom expression and with linear PSTV synthesis prevailing at later stages of the infection process. To investigate this possibility, PSTV-infected plants were allowed to incorporate ^{32}P beginning 1 week after the first day of symptom appearance. With a short period of incorporation (4 hours) labeling of circular PSTV was dominant, and with increasing duration of incorporation, the ratio of circular to linear labeled PSTV again decreased (Hadidi and Diener, 1978). These results clearly are not compatible with independent synthesis of circular and linear PSTV. They also rule out another hypothesis, namely that the level of RNase activity in infected plants at early stages of infection could be much lower than at later stages, thereby leading, during extraction, to the cleavage of fewer circular molecules at early stages of infection than at later stages.

In light of these results, Hadidi and Diener (1978) concluded that circular and linear PSTV molecules are two forms of the same RNA and that linear PSTV arises *in vivo* by cleavage of circular PSTV molecules.

9.5 MODE(S) OF PATHOGENESIS

No concrete evidence exists on how viroids interfere with their hosts'
metabolism to produce the characteristic macroscopic symptoms observed
in certain plant species and described in Chapter 2. In addition to these,
viroids may also cause more subtle aberrations at the cellular level. What
is known of these is summarized below. Possible molecular mechanisms of
pathogenesis are then considered.

9.5.1 Cytopathic Effects

Light microscopic comparison of sections made from healthy and PSTV-
infected leaves and petioles revealed a marked hypertrophy of nuclei in
the infected tissue. Also, protoplasmic streaming was found to be consistently more active in cells from infected than in those from healthy tissue
(J. F. Worley, personal communication, quoted in Diener, 1971c). Aside
from these changes, no clearly observable cytopathic effects were noted.

Semancik and Vanderwoude (1976) examined thin sections from CEV-
infected *Gynura* half leaves exhibiting symptoms and from opposite,
symptomless half-leaves by electron microscopy. They identified aberrations of the plasma membrane as the primary cytopathic effect associated
with CEV infection. The frequency of paramural bodies or plasmalem-
masomelike structures was reported to be directly correlated with the initiation of symptoms, as well as with the recovery of viroid (Semancik and
Vanderwoude, 1976). The authors suggested that alterations in cell-surface
properties may constitute a significant phase in viroid replication or pathogenesis.

Apparently, no other studies of cytopathic effects caused by viroid infection have been made.

9.5.2 Possible Mechanisms of Pathogenesis

The nuclear location and replication of viroids and their apparent inability to serve as messenger RNAs suggest that disease symptoms are caused
by interference of viroids with gene regulation in the infected host cells
(Diener, 1971b). It must be remembered, however, that some viroids have
extended host ranges (see Chapter 3) and that in the overwhelming majority of these host species no damage resulting from viroid infection is discernible. Thus in these symptomless hosts, viroid-induced metabolic aber-

rations do not apparently occur, or if they do occur, they must be harmless in the particular genetic milieu of the host.

As discussed in Section 9.2.1.2, with both PSTV and CEV, certain host proteins occur in larger amounts in infected than in healthy tissue. Possibly, these aberrations in host protein synthesis are related to the pathogenic properties of viroids.

If one accepts the postulate that viroids act as activator RNAs in Britten and Davidson's (1969) sense, explanation of their pathogenic properties poses little theoretical difficulty. Introduction of a viroid into a susceptible cell may result in the activation of previously silent producer genes and, in certain genetic milieus, this activation could lead to metabolic aberrations and ultimately to discernible macroscopic symptoms. However, no novel protein species has been detected in viroid-infected plants. It is more likely, therefore, that the regulating effect of viroids is a subtle quantitative one, resulting in overproduction of certain proteins, as has been observed. In the version of the Britten and Davidson model that assumes redundancy in receptor genes, a producer gene might normally be derepressed to a limited extent by nonviroid activator RNA complexed to one associated receptor gene; introducing of viroid could then lead to a complex with another associated receptor gene, and further activation of the producer gene might occur.

10. QUESTION OF ANIMAL VIROIDS

All viroids identified so far have been isolated from higher plants. The question arises as to whether similar entities exist in other forms of life. Above all, we wish to know whether viroids or viroidlike nucleic acids occur in mammals, including man, and whether such nucleic acids might be responsible for certain diseases of unknown etiology.

In view of the uniform biochemical basis of all life on earth, it is difficult to understand why viroidlike nucleic acids should exist only in higher plants. Yet, so far, no RNAs with the unique structural properties of plant viroids have been reported to occur in animal cells and no diseases of animals are definitely known to be caused by viroidlike nucleic acids.

The etiological agents of one group of animal diseases, however, possess certain properties suggesting similarity with plant viroids. These are the so-called "slow virus infections caused by unconventional viruses" (Gajdusek, 1977). The group consists of four known natural diseases: scrapie in sheep and goats (Hunter, 1974), transmissible mink encephalopathy (Marsh and Hanson, 1969), and, in man, Kuru (Gajdusek, 1977) and Creutzfeldt-Jakob disease (Gibbs et al., 1968). All four diseases have a long incubation period and cause similar brain lesions that include grey matter vacuolation and astrocyte hypertrophy. Also, they are characterized by the absence of a specific immune response of the host (Kimberlin, 1976).

Because, until recently, only primate hosts have been available as indicators for the agents causing disease in humans, and because of their very long incubation periods in their natural hosts, knowledge of the properties of the causative agents is mostly based on the study of the scrapie agent adapted to mice (Chandler, 1961; Hunter et al., 1976; Kimberlin, 1976) and hamsters (Kimberlin and Marsh, 1975; Marsh and Kimberlin, 1975).

Some of the most atypical properties of the scrapie agent are its high resistance to ultraviolet and ionizing radiations, as well as to various physical and chemical agents (such as heat, ultrasonic energy, formaldehyde, proteases, and nucleases), the absence of virionlike structures detectable by electron microscopy, the lack of inclusion bodies or demonstrable nonhost proteins in infected cells, and the scrapie agent's inability to elicit a specific immune response from its hosts (Gajdusek, 1977).

The remarkable resistance of the scrapie agent to ultraviolet and ionizing radiations has led some investigators to suggest that the agent does not depend on a nucleic acid replicating system. From the dose of ionizing radiation required to give an average of one inactivating event per infective unit of scrapie, Alper et al. (1966) calculated a "target volume" of 1.5×10^5. Because this value is much smaller than that for any known virus, the authors reasoned that if the "target" were nucleic acid, this molecular weight would be too low to allow of sufficient coding information for replication (Alper et al., 1966; Latarjet et al., 1970). Similarly, on the basis of its high resistance to ultraviolet radiation, Alper et al. (1967) concluded that scrapie was most unlikely to depend on a nucleic acid moiety for its replicative ability. Futhermore, Latarjet et al. (1970) demonstrated that the ultraviolet inactivation spectrum of scrapie was unusual for nucleic acid in that higher sensitivity was observed at 230 nm than either at 250 or 280 nm.

Because of these atypical properties of the scrapie agent, a number of ingenious, albeit unorthodox, hypotheses have been advanced to explain its apparent replication without the need of nucleic acid involvement. Thus Field (1966) suggested that the scrapie agent might be a replicating polysaccharide, whereas Pattison and Jones (1967) hypothesized that the agent might be, or might be associated with, a small basic protein. Gibbons and Hunter (1967), on the other hand, suggested that scrapie arose from a replicable change in the structural pattern of a commonly occurring unit membrane. Later, Hunter et al. (1968) advanced a modification of this membrane hypothesis, but Adams and Caspary (1967) maintained that such radical departures from conventional theory were not essential to explain the many exceptional properties of the scrapie agent and proposed that it might consist of a small nucleic acid core enclosed in a mucopolysaccharide or polysaccharide.

With the recognition of the small size of plant viroids (Diener, 1971b), it became apparent that the small target volume of the scrapie agent did not necessarily rule out nucleic acid as the genetic material. Because

PSTV, with a molecular weight corresponding to about 360 nucleotides, is capable of autonomous replication, no reason exists why a putative scrapie nucleic acid should not also be capable of autonomous replication.

Diener (1972c) pointed out many similarities between properties of PSTV and those of the scrapie agent and advanced the hypothesis that the scrapie agent might be a viroid. In support of this contention, the following similarities were pointed out: (1) Both scrapie and potato spindle tuber are infectious diseases, the agents of which greatly increase in inoculated hosts. Both diseases are characterized by long incubation periods.[1] (2) Electron microscopy of thin sections from tissue infected with either agent failed to disclose the presence of viruslike particles. (3) Both agents are unusually resistant to ionizing and ultraviolet radiation (see above, and Section 7.1.3). (4) The remarkable resistance of the scrapie agent to heating would not be surprising if the agent were a nucleic acid. As pointed out in Section 7.3.1, viroids tolerate heating as well if not better than the scrapie agent. (5) Neither agent exists in the soluble fraction of tissue homogenates, but is associated with subcellular particles (Diener, 1972c). The author recognized, however, that in several other properties the scrapie agent seems to differ fundamentally from PSTV. Thus the scrapie agent is insensitive to treatment with RNase or DNase (Hunter and Millson, 1967), whereas PSTV is sensitive to RNase. Also, the scrapie agent seems to be markedly unstable in the presence of strong phenol solutions (Hunter et al., 1969b), whereas the infectivity of PSTV in unaffected (Diener and Raymer, 1969) or enhanced (Singh and Bagnall, 1968). Furthermore, scrapie loses infectivity when exposed to cesium chloride (Mould et al., 1965), to high urea concentrations (Hunter et al., 1969b), or to ether (Gibbons and Hunter, 1967), whereas PSTV is stable in the presence of these compounds (Diener and Raymer, 1969). Diener (1972c) pointed out, however, that almost all of the stability tests with scrapie were made with crude brain homogenates and that the results, therefore, do not necessarily reflect properties of the scrapie agent per se. A basic disparity appears to exist also regarding the subcellular location of the respective agents. Whereas PSTV is mostly associated with nuclei, scrapie is known to be associated with plasma membranes (Kimberlin, 1976). This disparity, in fact, has been pointed out to discount the viroid hypothesis for scrapie (Gajdusek, 1973). But, as discussed in Section 9.3,

---

[1] With cadang-cadang disease of coconut palms, Randles et al. (1977) showed that the incubation period lasted from 1.5 to 2.0 years.

the possibility exists that some viroid may similarly be associated with plasma-membranelike components (Semancik et al., 1976).

Even less compatible with a viroid hypothesis appears to be the observation by several workers that highly purified nucleic acid preparations from scrapie-infected tissues are devoid of infectivity (Kimberlin, 1976). Three studies that were made with the viroid model in mind resulted in noninfectious nucleic acid preparations (Marsh et al., 1974; Ward et al., 1974; Hunter et al., 1976). In these studies, appropriate controls were used to show that the extraction procedures were capable of yielding functional nucleic acids.

Neither study, however, conclusively ruled out the possibility that the nucleic acid preparations from scrapie-infected brain did contain infectious nucleic acid, but that this infectivity was destroyed during intracerebral inoculation by exposure to serum (or other) nucleases. The complete insensitivity of the scrapie agent to nucleases in crude extracts documents that, if the agent is composed of, or contains, nucleic acid, this nucleic acid is well protected from nucleases. Thus inoculation with purified nucleic acid, as compared with crude extracts, can be expected to lead to a titer loss similar to that experienced with free infectious viral nucleic acids as compared with complete virus particles.

Indeed, in the study of Ward et al. (1974), intracerebral inoculation of intact mengovirus resulted in $10^{8.9}$ LD_{50} units per ml, whereas incoulation with phenol-liberated mengovirus RNA resulted in $10^{2.7}$ LD_{50} units per ml. Thus the titer loss resulting from inoculation with free viral RNA instead of intact virus amounted to 6.2 log units. Any putative scrapie nucleic acid could be expected to undergo a similar titer loss on intracerebral inoculation after having been liberated from protective components. Because the stock scrapie preparation used contained only $10^{5.9}$ mean lethal doses per ml (Ward et al., 1974), it is not surprising that no infectivity could be demonstrated in the purified nucleic acid preparations. Marsh et al. (1974) monitored their ability to extract and measure infectious nucleic acid by using encephalomyocarditis virus, but unfortunately they did not present data to indicate the titer loss experienced by intracerebral inoculation with free viral RNA as compared with the same amount of RNA in intact virus particles. Their scrapie (and transmissible mink encephalopathy) preparations, however, had titers of $10^{7.2}$ to $10^{7.5}$ LD_{50}/ml. Thus with a similar loss as that reported for mengovirus RNA by Ward et al. (1974), Marsh et al. (1974) should have been able to detect infectious scrapie (and mink encephalopathy) nucleic acids. The fact that they did not indicates that either no infectious nucleic acid was present or that larger titer losses than with mengovirus were experienced. This latter possibility is not unreasonable in view of the fact that with certain other viruses, such as lactic dehydrogenase virus (Notkins, 1965), the titer of infectious nucleic acid may be as much as eight logarithms less than the titer of the original intact virus.

Evidently, no concrete evidence exists to indicate that the unconventional slow infections of animals are caused by agents similar to plant viroids, but neither is there conclusive evidence that this is not the case. On the contrary, most scrapie investigators today accept the idea that the scrapie agent consists of, or contains, nucleic acid (Kimberlin, 1976). If so, in view of the irradiation studies, this nucleic acid must be of low molecular weight. Thus similarity of scrapie with plant viroids is a distinct possibility. Kimberlin (1976), in fact, advanced a model for the scrapie agent in which the scrapie specific information is assumed to be encoded in nucleic acid of 10^5 daltons which, however, is able to express full biological activity only when it is closely associated with some host membrane components to give a larger complex.

If this, or some similar, model proves to be correct, the difference between scrapielike agents and plant viroids may have to do more with the cellular components the respective agents are associated with *in situ* than with the nature of the genetic material which, in either case, would consist of low molecular weight nucleic acid.

Very small amounts of a single-stranded DNA have, indeed, been isolated from scrapie-infected but not from similarly prepared healthy tissue (Adams, 1972; Hunter *et al.*, 1973; Corp and Somerville, 1976). This DNA has an apparent molecular weight of 60,000, but its relationship, if any, with the scrapie agent has not yet been elucidated.

Evidence for an essential DNA component in the scrapie agent has recently been obtained with preparations from scrapie-infected hamster brains (Marsh *et al.*, 1978). Fractions derived from high-speed supernatants of brain homogenates that were further partitioned by either hydroxyapatite chromatography or polyacrylamide gel electrophoresis were incubated with DNase I, RNase A, or proteinase K. Subsequent bioassay demonstrated that incubation with RNase or proteinase K had no significant effect on infectivity titer, but that incubation with DNase I led to a titer loss of 1.5 to 3.3 log units.

The exact nature, size, and possible association of this scrapie-specific

DNA with other macromolecules, however, remains to be determined (Marsh *et al.*, 1978).

Many animal diseases of unknown etiology exist aside from the ones discussed so far, but whether any of these might be due to viroidlike agents is unknown, and in the complete absence of evidence for or against such a hypothesis, speculation seems unproductive.

11. NATURE OF VIROIDS

In contrast to the preceding chapters, in which factual knowledge of viroids was discussed, this chapter is on far less solid ground. We will ask ourselves fundamental questions, such as: What is the basic nature of viroids? Are there entities analogous or homologous to viroids in nature? What is the origin of viroids? In light of present knowledge, no definitive answers are possible to these questions; thus, by necessity, much of the following is speculative and tentative.

11.1 POSSIBLE HOMOLOGUES OR ANALOGUES OF VIROIDS

Viroids are the only species of single-stranded RNA known to exist both as linear molecules and as covalently closed circles. Thus exact homologues of viroids are unknown.

Certain cellular RNA species, however, possess structural properties that are somewhat similar to those of viroids. Seemingly double-stranded RNA has been shown to occur in apparently uninfected animal cells (Montagnier, 1968; Duesberg and Colby, 1969; Stern and Friedman, 1970; Stollar and Stollar, 1970). These RNA species are of low molecular weight (sedimenting at 4 to 14 S) and, although they have many of the attributes of double-stranded RNA, they are not simple base-paired structures (Stern and Friedman, 1970). These RNAs originate in the nucleus (Jelinek and Darnell, 1972) and hybridize rapidly to cellular DNA. They therefore appear to be transcribed from reiterated DNA sequences (Reanney, 1975). Similarity of these "double-stranded" RNAs with viroids has been pointed out and the question has been raised as to whether they might represent the normal counterparts of viroids (Diener, 1971b). In the absence of more detailed structural knowledge of these RNAs, however, such assumptions are questionable. Also, the functional significance of normally occurring "double-stranded" RNA is unknown.

Other cellular RNAs that structurally have some similarity with viroids are certain low molecular weight nuclear RNAs. One of these, 4.5 S RNA I from Novikoff hepatoma cell nuclei, has been sequenced (Ro-Choi *et al.*, 1972) and shown to have a very high degree of intramolecular self-complementarity. Also, this RNA is limited to the chromatin or nucleoplasm and, like viroids, is not found in any of the cytoplasmic fractions (Ro-Choi and Busch, 1974). Containing only 96 nucleotides, however, the 4.5 S RNA is far smaller than viroids. A number of other low molecular weight nuclear RNAs are known (Ro-Choi and Busch, 1974), but, here again, nothing is known of their possible function(s). Detailed discussion of these RNA species falls outside the scope of this book.

Low molecular weight RNAs associated with various viruses have also been reported, but detailed structural properties, to permit comparison with viroids, have been reported for only a few.

A low molecular weight RNA associated with adenovirus 2-infected KB cells, VA-RNA, has been sequenced and shown to have the potential for extensive regions of base-pairing (Ohe and Weissman, 1971). The function of this RNA, which is transcribed from adenovirus DNA (Ohe, 1972) and which is synthesized in large amounts late in the lytic cycle, is unknown (Philipson and Lindberg, 1974).

Low molecular weight RNAs are also associated with a number of plant viruses, such as certain strains of cucumber mosaic virus (Kaper *et al.*, 1976) and tobacco streak virus (Clark and Lister, 1971) among others, but neither structural nor functional similarities with viroids are apparent. A low molecular weight RNA that is associated with tobacco mosaic virus infection (LMC RNA) has been shown to contain the code for capsid protein (Siegel *et al.*, 1976).

A small RNA (220 nucleotides) that can be isolated from *in vitro* reaction mixtures containing bacteriophage Qβ replicase but no exogenous template RNA (Kacian *et al.*, 1972) has some properties resembling those of viroids. The complete nucleotide sequence of the RNA (MDV-1 RNA) has been determined (Mills *et al.*, 1973) and shown to permit extensive secondary and tertiary structures containing antiparallel stems and loops. Later, a still smaller (114 nucleotides) replicating molecule, microvariant RNA, has been isolated from similar reaction mixtures and its nucleotide sequence determined (Mills *et al.*, 1975). It is believed that these small RNAs originated in the *in vitro* systems as a consequence of selection pressures that favored rapid replication of Qβ RNA. Under these conditions, sequences that code for the coat proteins and replicase components are dispensable and are, therefore, eliminated (Mills *et al.*, 1967). The resulting RNAs may contain only sequences, such as those involved in the recognition mechanism employed by the replicase, that are required for their replication in the system provided. It appears permissible to regard these RNAs as artificially created counterparts of the naturally occurring viroids, because both types of molecules represent minimal self-duplicating entities that are entirely dependent for their replication on extraneously provided enzyme systems. In one case, the necessary systems are provided in the test tube, in the other by the invaded host cell.

Another approach to the question of viroid counterparts is to search for normal RNAs with functions similar to those of viroids. In this endeavor, we are stymied by our lack of knowledge of the mechanisms of *in vivo* viroid function.

Nevertheless, in view of the apparent inability of viroids to act as messenger RNAs, it seems plausible to assume, as has been proposed earlier (Diener, 1971b), that viroids represent abnormal regulatory RNAs; that is, that they interfere with gene regulation in the infected host cells. Whether this interference involves specific interaction with a nuclear regulatory protein or whether the viroid acts directly as an activator RNA in Britten and Davidson's (1969) sense is unknown, but as noted (McClements and Kaesberg, 1977), the hairpinlike structure of viroids provides the proper configuration for specific interaction of the RNA with a protein (Gralla *et al.*, 1974).

With this model in mind, it is interesting to examine the present status of the hypothesis that, in eukaryotic cells, specific RNA species may be directly involved in regulating gene expression and cell differentiation. Although both protein and RNA have been proposed as regulatory molecules affecting the pattern of gene activity in eukaryotic cells, RNA species with possible involvement in gene regulation have been described in recent years (Kanehisa *et al.*, 1971, 1972, 1974, 1977a, b; Goldstein, 1976). The concept of regulatory RNA has become more plausible with the isolation and characterization of a low molecular weight RNA species from embryonic chick heart that is capable of inducing a specific mode of changes in early embryonic cells of chick blastoderms cultivated *in vitro* (Deshpande *et al.*, 1977). This RNA seems to be an authentic component of the chick embryonic heart and is apparently absent in other chick embryonic tissues examined. The RNA sediments at about 7 S; it contains poly (A), yet was found to be nontranslatable in *in vitro* wheat germ and rabbit reticulocyte systems (Deshpande *et al.*, 1977). Another recent report describes two low molecular weight RNAs with possible regulatory functions in the embryos of *Artemia salina* (Lee-Huang *et al.*, 1977). One of these RNAs is a translational inhibitor of about 6000 daltons in size; the other is an activator RNA able to complex with the inhibitor to neutralize its inhibitory activity. Consequently, the second RNA can play a regulatory role causing the onset of protein synthesis in the developing embryo (Lee-Huang *et al.*, 1977).

Thus, several RNAs with possible involvement in the control of translation and transcription have emerged during recent years. No obvious structural similarities between these cellular RNAs and viroids are evident but the existence of low molecular weight cellular RNA species with apparent regulatory roles makes it more plausible that viroids may similarly interfere with gene regulation in their hosts.

Analogies might be drawn also between viroids and other cellular entities, but none of these appear as plausible as comparison with putative regulatory RNAs. Viroids might be compared with plasmids, for example, but these usually are cytoplasmic constituents and composed of DNA, although a trait of the yeast *Saccharomyces cerevisiae* that displays a non-Mendelian pattern of heredity has been correlated with the presence of an RNA plasmid (Wickner, 1976). Similarities with viroids, however, are not apparent because the RNA in yeast is double stranded, has a molecular weight of 1.4 to 1.7×10^6, and is encapsulated in viruslike particles (Wickner, 1976).

Functionally, viroids might be compared to transfer factors or insertion sequences, but in view of the basic chemical differences such comparisons do not appear to be productive.

Recent studies in many laboratories [for reviews, see Sambrook (1977); Williams (1977); Gilbert (1978)] suggest that the coding sequences on the DNAs of eukaryotic organisms (as well as of certain viruses) in general are not continuous but are interrupted by "silent" DNA and that the primary RNA transcripts contain internal regions that are excised during maturation. Thus the final messenger RNA (or even tRNA) is a spliced product. Gilbert (1978) proposed the terms *intron,* to denote those regions which will be lost from the mature messenger, and *exon*, to denote the expressed regions. The eukaryotic gene thus is a mosaic consisting of expressed sequences held in a matrix of "silent" DNA, an intronic matrix (Gilbert, 1978). The introns so far identified range from 10 to 10,000 bases in length. This concept readily explains the existence of heterogeneous nuclear RNA which clearly is the transcription product out of which the much smaller ultimate messengers for expressed protein sequences are spliced (Gilbert, 1978). Evidently, this splicing process requires the presence of RNA ligases to reseal the cuts. Whether the spliced-out sections of heterogeneous nuclear RNAs fulfill a cellular function (such as playing a role in gene regulation) or whether they simply constitute excess RNA that is rapidly degraded has not, apparently, been established.

Such an intramolecular splicing process, however, could explain the formation of viroidlike, circular RNA molecules. It is conceivable that, because of their great stability, spliced-out sections (introns) that happen to have nucleotide sequences permitting extensive intramolecular base-pairing may escape degradation and be ligated. Such molecules evidently would possess viroidlike structures; they might be replicated by host enzymes capable of acting as RNA-directed RNA polymerases and thereby escape the regulatory control of the cell.

11.2 POSSIBLE ORIGIN OF VIROIDS

Before the unique molecular structure of viroids was recognized, it was not unreasonable to consider the possibility that viroids might have originated from conventional viruses by degeneration (Diener, 1974). Alternatively, it was thought, viroids might represent primitive viruses that had not yet developed the genetic sophistication to code for one or more capsid proteins capable of assembling into a protective capsid.

With the demonstration of their unique molecular structure and of host DNA sequences complementary to the viroid, a virus-derived origin of viroids has become unlikely.

Far more plausible, in light of present knowledge, is the contention that viroids originated from normal cellular RNAs of their hosts, and it is tempting to speculate that they specifically originated from nuclear low molecular weight RNAs that perform a normal regulatory function. The change from a normal constituent to a pathogenic RNA might have occurred either by mutation or by chance introduction into a foreign species, in which the RNA is replicated. In either case, the disease would be a consequence of interference with the functions of normally occurring nuclear RNAs (Diener, 1974). These hypotheses are further developed in the following section.

11.3 SPECULATIONS ON THE NATURE OF VIROID DISEASES

In contrast to many diseases of cultivated plants caused by conventional viruses, the diseases now known to be viroid incited have come to the attention of plant pathologists only recently. Thus, as discussed in Chapter 2, the first report of potato spindle tuber disease goes back only to 1922, of tomato bunchy top to 1931, of citrus exocortis to 1948, of chrysanthemum stunt to 1947, of chrysanthemum chlorotic mottle to 1969, of cucumber pale fruit to 1974, and of coconut cadang-cadang to 1937. One wonders whether the diseases in question had been present previously for a long time and simply were not recognized or whether they developed as agricultural problems in the recent past only. The former alternative is possible in the case of potato spindle tuber disease which, according to Schultz and Folsom (1923a), had been recognized by growers for many years and constituted one aspect of the then illdefined "running out" problem of cultivated potato. Citrus exocortis disease, similarly, was probably present as far back as the early 1920s in California and South Africa (see Section 2.3.1).

With some other viroid diseases, however, it is far more difficult to maintain that they have existed for a long time and simply have been overlooked by growers and plant pathologists alike. Thus chrysanthemum stunt disease has only been recognized since 1945 (Dimock, 1947). It appears unlikely that the disease could have long gone unrecognized because, only one year later in 1946, the disease had become generally prevalent in the U.S.A. and Canada (Brierley and Smith, 1949). Chrysanthemum chlorotic mottle disease was first seen in 1967 and described as a "new" disease 2 years later (Dimock and Geissinger, 1969). Equally brief is the recorded history of cucumber pale fruit disease which was first observed in 1963 in two glasshouses in the western part of the Netherlands and which has since been observed in different places over the whole country (Van Dorst and Peters, 1974). It seems inconceivable that, in a country with as sophisticated and extensive a plant pathological organization as the Netherlands, the disease could have long escaped attention. Similar considerations apply to coconut cadang-cadang disease whose first appearance in one area of one small island of the Philippines has been documented (see Section 2.7.1). Again, once the disease had been recognized its spread was rapid, and it appears therefore unlikely that the disease could have existed unrecognized for a long time on a few palm trees without sufficient spread for the condition to become recognized. Also, because the disease has not been reported to occur elsewhere, it is difficult to accept the idea that it was imported from abroad.

In light of these observations, it appears reasonable to postulate that *viroid diseases of cultivated plants are of recent origin.*

In fact, none of the presently recognized viroid diseases has been known to exist before the Twentieth Century. Why did they appear now and not in earlier centuries? It seems likely that some aspect of modern agricultural practice has favored their appearance. We may thus postulate that *viroid diseases of cultivated plants are caused by man and his agricultural activities.*

Indeed, as discussed in Chapter 2, no reliable reports of insect vectors of viroids exist; the only known vector seems to be man and his tools. Ecologically, viroid diseases have become possible only, so it seems, with the introduction of modern intensive methods of agriculture. What particular aspects are involved is unknown, but one suspects that the widespread practice of monoculture of crop plants is involved.

Although these ideas may explain why viroid diseases are a recent phenomenon, they do not account for the source of the pathogens involved. Conceivably, viroids could have originated in the cultivated plant species themselves—for example, by mutation of normal cellular RNAs. If so, the recent origin of viroid diseases is difficult to understand. Far more plausible, it appears, is the assumption that viroid reservoirs exist in species of wild plants and that viroid diseases originate by chance transfer of a viroid from a wild carrier species to a susceptible cultivated plant species. We may postulate that *viroid diseases usually originate by accidental introduction of viroids from reservoir plants into susceptible cultivated plants.*

This postulate is supported by the observation that species closely related to the cultivated potato, such as *Solanum phureja* and *S. stenotonum,* frequently harbor viroids capable of inducing potato spindle tuber disease in cultivated potato (Owens *et al.,* 1978). Also, seemingly healthy ornamental plants (*Columnea erythrophae*) have been shown to often be viroid infected. Undoubtedly, future work will unearth many more instances of viroids, capable of causing disease in certain cultivated plants, being harbored by seemingly healthy individuals of other plant species. It is of interest to note in this connection that the experimental host ranges of several viroids include many wild species (Tables 1 to 5) and that most of these species tolerate viroid replication without the appearance of recognizable disease symptoms. Thus, as a rule, viroids cause disease mainly in cultivated plant species and only rarely in wild species (presumably their natural hosts). In view of these observations, we may postulate that *viroids often are latent in their natural hosts but may become pathogenic only when transferred to other species.*

We may carry these speculations one step further by asking ourselves *why* viroids are latent in what are here assumed to be their natural hosts, whereas they are often pathogenic in cultivated plant species.

One possible explanation starts with the postulate that *viroids evolved in their natural hosts from normal regulatory RNAs* (see Section 11.1). Because putative regulatory RNAs must be stable enough to resist degradation in their passage to target DNA, they may be expected to possess a high degree of intramolecular self-complementarity (Reanney, 1975). For RNA to be able not only to survive intracellular passage to target DNA, but passage from cell to cell and even from organism to organism, still more stringent stability requirements would have to be met. In this model, then, those regulatory RNAs that evolved extremely stable secondary structures, that is, structures permitting extensive regions of base-pairing, were able to escape the confines of their cells of origin and thereby acquired the potential to become viroids. We postulate that *viroids originated from normal regulatory RNAs that evolved into structures of high stability and thus acquired the capacity for intercellular and interorganismal mobility.*

This model implies that viroid progenitors are transcribed from host DNA but leaves open the possibility that, in their newly acquired host species, viroid replication may be DNA independent. In these latter species, host enzymes capable of functioning as RNA-directed polymerases (see Section 9.4.3.1) may use the infecting viroid molecule as a template for synthesis of complementary RNA strands from which progeny viroid molecules are then transcribed.

The model accounts for the observation that viroids often are latent in wild plant species (presumably their natural hosts), whereas they often are pathogenic in cultured plant species. In the former species, viroids may perform normal regulatory functions, whereas in the latter species with their different genetic milieus, the regulatory action of the introduced viroid may lead to metabolic aberrations and ultimately to macroscopically discernible disease symptoms.

Finally, the model makes two predictions:

1. If its postulates are essentially correct, new viroid diseases of cultivated plants will continue to develop. They will seemingly appear from nowhere.

2. Viroidlike RNAs will be found in many more apparently healthy plants of diverse species. Some of these RNAs will be pathogenic in certain cultivated plant species.

It will be interesting to see whether future work will substantiate or refute the several postulates of this speculative model of viroid nature.

REFERENCES

Adams, D. H. (1970). The nature of the scrapie agent. A review of recent progress. *Path. Biol.*, **18**, 559–577.

Adams, D. H. (1972). Studies on DNA from normal and scrapie-affected mouse brain. *J. Neurochem.*, **19**, 1869–1882.

Adams, D. H., and Caspary, E. A. (1967). Nature of the scrapie virus. *Brit. Med. J.*, *iii*, 173.

Akeley, R. V., Stevenson, F. J., Schultz, E. S., Bonde, R., Neilsen, K. F., and Hawkins, A. (1955). Saco, a new late-maturing variety of potato, immune from common race of the late blight fungus, highly resistant to if not immune from net necrosis, and immune from mild and latent mosaics. *Am. Potato J.*, **32**, 41–48.

Allen, R. M., (1968). Survival time of exocortis virus of citrus on contaminated knife blades. *Plant Dis. Reporter*, **52**, 935–939.

Allington, W. B., Ball, E. M., and Galvez, G. (1964). Potato spindle tuber caused by a strain of potato virus X. *Plant Dis. Reporter*, **48**, 597–598.

Alper, T., Haig, D. A., and Clarke, M. C. (1966). The exceptionally small size of the scrapie agent. *Biochem. Biophys. Res. Commun.*, **22**, 278–284.

Alper, T., Cramp, W. A., Haig, D. A., and Clarke, M. C. (1967). Does the agent of scrapie replicate without nucleic acid? *Nature*, **214**, 764–766.

Altenburg, E. (1946). The "viroid" theory in relation to plasmagenes, viruses, cancer, and plastids. *Am. Naturalist*, **80**, 559–567.

Aoki, S., and Takebe. I. (1969). Infection of tobacco mesophyll protoplast by tobacco mosaic virus ribonucleic acid. *Virology*, **39**, 439–448.

Astier-Manifacier, S., and Cornuet, P. (1971). RNA-dependent RNA polymerase in Chinese cabbage. *Biochim. Biophys. Acta*, **232**, 484–493.

Babos, P., and Shearer, G. B. (1969). RNA synthesis in tobacco leaves infected with tobacco mosaic virus. *Virology*, **39**, 286–295.

Bagnall, R. H. (1967). Serology of the potato spindle tuber virus. *Phytopathology*, **57**, 533–534.

Bagnall, R. H., Larson, R. H., and Walker, J. C. (1956). Potato viruses M, S, and X in relation to interveinal mosaic of the Irish Cobbler variety. *Wisconsin Agr. Exp. Sta. Res. Bull.*, **198**, 45 pp.

Ball, E. M., Allington, W. B., and Galvez, G. E. (1964). Serological studies of a virus that produced spindle tuber in potatoes. *Phytopathology*, **54**, 887.

Bancroft, J. B., and Key, J. L. (1964). Effect of actinomycin D and ethylenediamine tetraacetic acid in the multiplication of a plant virus in etiolated soybean hypocotyls. *Nature*, **202**, 729–730.

Bennett, C. W. (1969). Seed transmission of plant viruses. *Adv. Virus Res.*, **14**, 221–261.

Benson, A. P., and Singh, R. P. (1964). Seed transmission of potato spindle tuber virus in tomato. *Am. Potato J.*, **41**, 294.

Benson, A. P., Salama, F. M., and Singh, R. P. (1964). Concentration and characterization of potato spindle tuber virus. *Am. Potato J.*, **41**, 293.

Benson, A. P., Raymer, W. B., Smith, W., Jones, E., and Munro, J. (1965). Potato diseases and their control. In: *Potato Handbook*, **10**, 32–36.

Benton, R. J., Bowman, F. T., Fraser, L., and Kebby, R. G. (1949). Stunting and scaly butt of citrus associated with *Poncirus trifoliata* rootstock. *Agr. Gaz. N.S. Wales*, **61**, 521–526, 577–582, 641–645, 654.

Benton, R. J., Bowman, F. T., Fraser, L., and Kebby, R. G. (1950). Stunting and scaly butt of citrus associated with *Poncirus trifoliata* rootstock. *Agr. Gaz. N.S. Wales,* **61**, 20–22, 40.

Billeter, M. A., Weissmann, C., and Warner, R. C. (1966). Replication of viral ribonucleic acid. IX. Properties of double-stranded RNA from *Escherichia coli* infected with bacteriophage MS 2. *J. Mol. Biol.*, **17**, 145–173.

Bitters, W. P. (1952). Exocortis disease of citrus. Top-root relationships of trifoliate orange and its hybrids studied in search for cause of root disease. *Calif. Agricult.*, **6**, 5–6.

Bitters, W. P., Brusca, J. A., and Dukeshire, N. W. (1954). Effect of lemon budwood selection in transmission of exocortis. *Citrus Leaves*, **34** (1), 8–9, 34.

Boedtker, H. (1971). Conformation independent molecular weight determinations of RNA by gel electrophoresis. *Biochim. Biophys. Acta*, **240**, 448–453.

Bol, J. F., Clerx-Van Haaster, C. M., and Weening, C. J. (1976). Host and virus specific RNA polymerases in alfalfa mosaic virus infected tobacco. *Ann. Microbiol. (Inst. Pasteur)*, **127A**, 183–192.

Bonde, R. (1927). The spread of spindle tuber by the knife. *Am. Potato J.*, **4**, 51–52.

Bonde, R., and Merriam, D. (1951). Studies on the dissemination of the potato spindle tuber virus by mechanical inoculation. *Am. Potato J.*, **28**, 558–560.

Brakke, M. K. (1970). Systemic infections for the assay of plant viruses. *Ann. Rev. Phytopath.*, **8**, 61–84.

Brierley, P. (1950). Some host plants of chrysanthemum stunt virus. *Phytopathology*, **40**, 869.

Brierley, P. (1952). Exceptional heat tolerance and some other properties of the chrysanthemum stunt virus. *Plant Dis. Reporter*, **36**, 243–244.

Brierley, P. (1953). Some experimental hosts of the chrysanthemum stunt virus. *Plant Dis. Reporter*, **37**, 343–345.

Brierley, P., and Olson, C. J. (1956). Development and production of virus-free chrysanthemum propagative ma*erial. *Plant Dis. Rep. Suppl.*, **238**, 63–67.

Brierley, P., and Smith, F. F. (1949). Chrysanthemum stunt. *Phytopathology*, **39**, 501.

Brierley, P., and Smith, F. F. (1951). Chrysanthemum stunt. *Flor. Rev.*, 107 (2778), 27–30.

Brierley, P., Parker, M. W., and Borthwick, H. A. (1952). High-intensity artificial light improves winter expression of stunt symptoms in Mistletoe chrysanthemums. *Phytopathology*, **42**, 341.

Britten, R. J., and Davidson, E. H. (1969). Gene regulation for higher cells: A theory. *Science,* **165,** 349–357.

Brownlee, G. G., and Sanger, F. (1969). Chromatography of ³²P-labeled oligonucleotides on thin layers of DEAE-cellulose. *Eur. J. Biochem.,* **11,** 395–399.

Brownlee, G. G., Sanger, F., and Barrell, B. G. (1968). The sequence of 5 S ribosomal ribonucleic acid. *J. Mol. Biol.,* **34,** 379–412.

Burnett, H. C. (1961). The color test for exocortis indexing in Florida. In: *Proc. 2nd Conference Int. Org. Citrus Virologists* (W. C. Price, Ed.), Univ. Florida Press, Gainesville, pp. 22–25.

Cadman, C. H. (1962). Evidence for association of tobacco rattle virus nucleic acid with a cell component. *Nature,* **193,** 49–52.

Calavan, E. C., and Weathers, L. G. (1959). The distribution of exocortis virus in California citrus. In: *Citrus Virus Diseases* (J. M. Wallace, Ed.), Univ. California Div. Agr. Sci., pp. 151–154.

Calavan, E. C., and Weathers, L. G. (1961). Evidence for strain differences and stunting with exocortis virus. In: *Proc. 2nd Conference Int. Org. Citrus Virologists* (W. C. Price, Ed.), Univ. Florida Press, Gainesville, pp. 26–33.

Calavan, E. C., Soost, R. K., and Cameron, J. W. (1959). Exocortis-like symptoms on unbudded seedlings and rootstocks of *Poncirus trifoliata* with seedling-line tops, and probable spread of exocortis in a nursery. *Plant Dis. Reporter,* **43,** 374–379.

Calavan, E. C., Frolich, E. F., Carpenter, J. B., Roistacher, C. N., and Christiansen, D. W. (1964). Rapid indexing for exocortis of citurs. *Phytopathology,* **54,** 1359–1362.

Calavan, E. C., Weathers, L. G., and Christiansen, D. W. (1968). Effect of exocortis on production and growth of Valencia orange trees on trifoliate orange rootstock. In: *Proc. 4th Conference Int. Org. Citrus Virologists* (J. F. L. Childs, Ed), Univ. Florida Press, Gainesville, pp. 101–104.

Cammack, R. H. (1964). The abnormality resembling potato spindle tuber in the variety Redskin. *Plant Pathology,* **13,** 69–72.

Cammack, R. H. and Richardson, D. E. (1963). Suspected potato spindle tuber virus in England. *Plant Pathology,* **12,** 23–26.

Celino, M. S. (1947a). A preliminary report on a blight disease (cadang-cadang) of coconut in San Miguel Estate, Albay Province. *Philippine J. Agr.,* **13,** 31–35.

Celino M. S. (1947b). Progress report on experimental transmission of cadang-cadang disease of coconut. *Philippine J. Agr.,* **13,** 109–113.

Chandler, R. L. (1961). Encephalopathy in mice produced with scrapie brain material. *Lancet,* **i,** 1378–1379.

Chester, K. S. (1937). Serological studies of plant viruses. *Phytopathology,* **27,** 903–912.

Childs, J. F. L., Norman, G. G., and Eichhorn, J. L. (1958). A color test for exocortis infection in *Pancirus trifoliata. Phytopathology,* **48,** 426–432.

Clark, M. F., and Lister, R. M. (1971). Preparations and some properties of the nucleic acid of tobacco streak virus. *Virology,* **45,** 61–74.

Commerford, S. L. (1971). Iodination of nucleic acids *in vitro. Biochemistry,* **10,** 1993–2000.

Conejero, V., and Semancik, J. S. (1977). Exocortis viroid: Alteration in the proteins of *Gynura aurantiaca* accompanying viroid infection. *Virology,* **77,** 221–232.

Corp, C. R., and Somerville, R. A. (1976). Nucleic acids associated with detergent-treated synaptosomal plasma membranes from normal and scrapie-infected mouse brain. *Biochem. Soc. Trans.*, **4**, 1112–1113.

Davies, J. W., Kaesberg, P., and Diener, T. O. (1974). Potato spindle tuber viroid. XII. An investigation of viroid RNA as a messenger for protein synthesis. *Virology*, **61**, 281–286.

Deshpande, A. K., Jakowlew, S. B., Arnold, H. H., Crawford, P. A., and Siddiqui, M. A. Q. (1977). A novel RNA affecting embryonic gene functions in early chick blastoderm. *J. Biol. Chem.*, **252**, 6521–6527.

Dickson, E. (1976). Studies of plant viroid RNA and other RNA species of unusual function. Ph.D. Thesis, The Rockefeller University, New York.

Dickson, E., Prensky, W., and Robertson, H. D. (1975). Comparative studies of two viroids: Analysis of potato spindle tuber and citrus exocortis viroids by RNA fingerprinting and polyacrylamide-gel electrophoresis. *Virology*, **68**, 309–316.

Dickson, E., Diener, T. O., and Robertson, H. D. (1978). Potato spindle tuber and citrus exocortis viroids undergo no major sequence changes during replication in two different hosts. *Proc. Natl. Acad. Sci. U.S.A.*, **75**, 951–954.

Diener, T. O. (1962). Isolation of an infectious, ribonuclease-sensitive fraction from tobacco leaves recently inoculated with tobacco mosaic virus. *Virology*, **16**, 140–146.

Diener, T. O. (1968). Potato spindle tuber virus: *in situ* sensitivity of the infectious agent to ribonuclease. *Phytopathology*, **58**, 1048.

Diener, T. O. (1970). Isolation of exonuclease-resistant ribonucleic acid from healthy and potato spindle tuber virus-infected tomato leaves. *Phytopathology*, **60**, 1014.

Diener, T. O. (1971a). Potato spindle tuber virus: A plant virus with properties of a free nucleic acid. III. Subcellular location of PSTV-RNA and the question of whether virions exist in extracts or *in situ*. *Virology*, **43**, 75–89.

Diener, T. O. (1971b). Potato spindle tuber "virus." IV. A replicating, low molecular weight RNA. *Virology*, **45**, 411–428.

Diener, T. O. (1971c). A plant virus with properties of a free ribonucleic acid: Potato spindle tuber virus. In: *Comparative Virology* (K. Maramorosch and E. Kurstak, Eds.), Academic Press, New York, pp. 433–478.

Diener, T. O. (1972a). Potato spindle tuber viroid. VIII. Correlation of infectivity with a UV-absorbing component and thermal denaturation properties of the RNA. *Virology*, **50**, 606–609.

Diener, T. O. (1972b). Viroids. *Adv. Virus Res.*, **17**, 295–313.

Diener, T. O. (1972c). Is the scrapie agent a viroid? *Nature New Biology*, **235**, 218–219.

Diener, T. O. (1973a). Virus terminology and the viroid: a rebuttal. *Phytopathology*, **63**, 1328–1329.

Diener, T. O. (1973b). A method for the purification and reconcentration of nucleic acids eluted or extracted from polyacrylamide gels. *Anal. Biochem.*, **55**, 317–320.

Diener, T. O. (1974). Viroids as prototypes or degeneration products of viruses. In: *Viruses, Evolution and Cancer* (E. Kurstak and K. Maramorosch, Eds.), Academic Press, New York, pp. 757–783.

Diener, T. O. (1977). Viroids: Autoinducing regulatory RNAs? In: *Genetic Interaction and Gene Transfer* (C. W. Anderson, Ed.), Brookhaven Symposium in Biology, No. 29, pp. 50–61.

Diener, T. O., and Hadidi, A. (1977). Viroids. In: *Comprehensive Virology* (H. Fraenkel-

Conrat and R. R. Wagner, Eds.), Plenum Publ. Corp., New York, Vol. 11, pp. 285–337.

Diener, T. O., and Heinze, R. G. (1962). Lesion formation on bean and *Datura stramonium* leaves rubbed with water or cupric sulfate and subsequently inoculated with tobacco mosaic virus nucleic acid. *Phytopathology*, **52**, 7.

Diener, T. O., and Lawson, R. H. (1972). Chrysanthemum stunt, a viroid disease. *Phytopathology*, **62**, 754.

Diener, T. O., and Lawson, R. H. (1973). Chrysanthemum stunt: A viroid disease. *Virology*, **51**, 94–101.

Diener, T. O., and Raymer, W. B. (1967). Potato spindle tuber virus: A plant virus with properties of a free nucleic acid. *Science*, **158**, 378–381.

Diener, T. O., and Raymer, W. B. (1969). Potato spindle tuber virus: A plant virus with properties of a free nucleic acid. II. Characterization and partial purification. *Virology*, **37**, 351–366.

Diener, T. O., and Raymer, W. B. (1971). *Potato spindle tuber "virus."* Commonwealth Mycological Inst./Assoc. Appl. Biol. Descript. Plant Viruses, No. 66. 4 pp.

Diener, T. O., and Smith, D. R. (1971). Potato spindle tuber viroid. VI. Monodisperse distribution after electrophoresis in 20% polyacrylamide gels. *Virology*, **46**, 498–499.

Diener, T. O., and Smith, D. R. (1973). Potato spindle tuber viroid. IX. Molecular-weight determination by gel electrophoresis of formylated RNA. *Virology*, **53**, 359–365.

Diener, T. O., and Smith, D. R. (1975). Potato spindle tuber viroid. XIII. Inhibition of replication by actinomycin D. *Virology*, **63**, 421–427.

Diener, T. O., and Weaver, M. L. (1959). Reversible and irreversible inhibition of necrotic ringspot virus in cucumbers by pancreatic ribonuclase. *Virology*, **7**, 419–427.

Diener, T. O., Smith, D. R., and O'Brien, M. J. (1972). Potato spindle tuber viroid. VII. Susceptibility of several solanaceous plant species to infection with low molecular-weight RNA. *Virology*, **48**, 844–846.

Diener, T. O., Schneider, I. R., and Smith, D. R. (1974). Potato spindle tuber viroid. XI. A comparison of the ultraviolet light sensitivities of PSTV, tobacco ringspot virus, and its satellite. *Virology*, **57**, 577–581.

Diener, T. O., Hadidi, A., and Owens, R. A. (1977). Methods for studying viroids. In: *Methods in Virology* (K. Maramorosch and H. Koprowski, Eds.). Vol. 6, pp. 185–217.

Dimock, A. W. 1947. Chrysanthemum stunt, *New York State Flower Growers Bull.*, **26**, 2.

Dimock, A. W., and Geissinger, C. (1969). A newly recognized disease of chrysanthemums caused by a graft-transmissible agent. *Phytopathology*, **59**, 1024.

Dimock, A. W., Geissinger, C. M., and Horst, R. K. (1971). Chlorotic mottle: A newly recognized disease of chrysanthemum. *Phytopathology*, **61**, 415–419.

Domdey, H., Jank, P., Sänger, H. L., and Gross, H. J. (1978). Studies on the primary and secondary structure of potato spindle tuber viroid: Products of digestion with ribonuclease A and ribonuclease T_1 and modification with bisulfite. *Nucleic Acids Res.*, **5**, 1221–1236.

Dorland, W. A. N. (1932). The American Illustrated Medical Dictionary. Sixteenth ed., W. B. Saunders Co., Philadelphia, Pa., 1493 pp.

Downey, K. M., Byrnes, J. J., Jurmark, B. S., and So, A. G. (1973). Reticulocyte RNA-dependent RNA polymerase. *Proc. Natl. Acad. Sci. U.S.A.*, **70**, 3400–3404.

Duda, C. T., Zaitlin, M., and Siegel, A. (1973). *In vitro* synthesis of double-stranded RNA by an enzyme system isolated from tobacco leaves. *Biochim. Biophys. Acta*, **319**, 62–71.

Duesberg, P. H., and Colby, C. (1969). On the biosynthesis and structure of double-stranded RNA in vaccinia-virus-infected cells. *Proc. Natl. Acad. Sci. U.S.A.*, **64**, 396–403.

Easton, G. D., and Merriam, D.C. (1963). Mechanical inoculation of the potato spindle tuber virus in the genus *Solanum. Phytopathology*, **53**, 349.

Engelhardt, D. L. (1972). Assay for secondary structure in ribonucleic acid. *J. Virol.*, **9**, 903–908.

Fawcett, H. S., and Klotz, L. J. (1948). Bark shelling of trifoliate orange. *The Calif. Citrograph*, **33**, 230.

Fernandez Valiela, M. V., and Calderon, A. V. (1965). The search for potato growing areas in the Argentine Republic. *Atlas Inst. Micol.*, **2**, 60–76 (quoted from *Rev. Appl. Mycol.*, **45**, 552, 1966).

Fernow, K. H. (1923). Spindling tuber or marginal leafroll. *Phytopathology*, **13**, 40.

Fernow, K. H. (1967). Tomato as a test plant for detecting mild strains of potato spindle tuber virus. *Phytopathology*, **57**, 1347–1352.

Fernow, K. H., Peterson, L. C., and Plaisted, R. L. (1969). The tomato test for eliminating spindle tuber from potato planting stock. *Am. Potato J.*, **46**, 424–429.

Fernow, K. H., Peterson, L. C., and Plaisted, R. L. (1970). Spindle tuber virus in seeds and pollen of infected potato plants. *Am. Potato J.*, **47**, 75–80.

Field, E. J. (1966). Transmission experiments with multiple sclerosis: An interim report. *Brit. Med. J.*, **ii**, 564–565.

Folsom, D. (1923). Potato spindle-tuber. *Maine Agr. Exp. Sta. Bull.*, **312**, 21–44.

Folsom, D. (1926). Virus diseases of the potato. *Quebec Soc. for the Protection of Plants, 18th Ann. Rpt.*, pp. 14–29 (quoted in Goss, 1930a).

Fraenkel-Conrat, H. (1956). The role of the nucleic acid in the reconstitution of active tobacco mosaic virus. *J. Am. Chem. Soc.*, **78**, 882–883.

Fraenkel-Conrat, H. (1976). RNA polymerase from tobacco necrosis virus infected and uninfected tobacco. Purification of the membrane-associated enzyme. *Virology*, **72**, 23–32.

Fraenkel-Conrat, H., Veldee, S., and Woo, J. (1964). The infectivity of tobacco mosaic virus. *Virology*, **22**, 432–433.

Francki, R. I. B. (1968). Inactivation of cucumber mosaic virus (Q strain) nucleoprotein by pancreatic ribonuclease. *Virology*, **34**, 694–700.

Franklin, R. M. (1966). Purification and properties of the replicative intermediate of the RNA bacteriophage R 17. *Proc. Natl. Acad. Sci. U.S.A.*, **55**, 1504–1511.

Fraser, L. R., and Levitt, E. C. (1959). Recent advances in the study of exocortis (scaly butt) in Australia. In: *Citrus Virus Diseases* (J. M. Wallace, Ed.), Univ. California Div. Agr. Sci., pp. 129–133.

Gajdusek, D. C. (1973). Kuru and Creutzfeld-Jakob disease. *Ann. Clin. Res.*, **5**, 254–261.

Gajdusek, D. C. (1977). Unconventional viruses and the origin and disappearance of Kuru. *Science*, **197**, 943–960.

Garnsey, S. M., and Jones, J. W. (1967). Mechanical transmission of exocortis virus with contaminated budding tools. *Plant Dis. Reporter*, **51**, 410–413.

Garnsey, S. M., and Whidden, R. (1970). Transmission of exocortis virus to various citrus plants by knife-cut inoculation. *Phytopathology*, **60**, 1292.

Garnsey, S. M., and Whidden, R. (1973). Efficiency of mechanical inoculation procedures for citrus exocortis virus. *Plant Dis. Reporter*, **57**, 886–890.

Geelen, J. L. M. C., Weathers, L. G., and Semancik, J. S. (1976). Properties of RNA polymerases of healthy and citrus exocortis viroid-infected *Gynura aurantiaca* DC. *Virology*, **69**, 539–546.

Gibbons, R. A., and Hunter, G. D, (1967). Nature of the scrapie agent. *Nature*, **215**, 1041–1043.

Gibbs, C. J., Jr., Gajdusek, D. C., Asher, D. M., Alpers, M. P., Beck, E., Daniel, P. M., and Matthews, W. B. (1968). Creutzfeld-Jakob disease (spongiform encephalopathy): Transmission to the chimpanzee. *Science*, **161**, 388–389.

Gierer, A. (1958). Grösse und Struktur der Ribonsenucleinsäure des Tabakmosaikvirus. *Z. Naturforsch. B*, **13**, 477–484.

Gierer, A. (1973). Molecular models and combinatorial principles in cell differentiation and morphogenesis. *Cold Spring Harbor Symp. Quant. Biol.*, **38**, 951–961.

Gierer, A., and Schramm, G. (1956). Infectivity of ribonucleic acid from tobacco mosaic virus. *Nature*, **177**, 702–703.

Gilbert, A. H. (1923). Spindling-tuber. *Vermont Agr. Ext. Service, College of Agr. Circular*, No. 28, 4 pp.

Gilbert, A. H. (1925). "Giant Hill" potatoes a dangerous source of seed. A new phase of spindle tuber. *Vermont Agr. Exp. Sta. Bull.*, **245**, 16 pp.

Gilbert, W. (1978). Why genes in pieces? *Nature*, **271**, 501.

Goldstein, L. (1976). Role for small nuclear RNAs in "programming" chromosomal information? *Nature*, **261**, 519–521.

Goss, R. W. (1924). Effect of environment on potato degeneration diseases. *Univ. Nebraska Agr. Exp. Sta. Res. Bull.*, **26**, 40 pp.

Goss, R. W. (1926a). Transmission of potato spindle-tuber by cutting knives and seed piece contact. *Phytopathology*, **16**, 299–304.

Goss, R. W. (1926b). Transmission of potato spindle-tuber disease by cutting-knives and seed piece contact. *Phytopathology*, **16**, 68–69.

Goss, R. W. (1926c). A simple method of inoculating potatoes with the spindle-tuber disease. *Phytopathology*, **16**, 233.

Goss, R. W. (1928). Transmission of potato spindle-tuber by grasshoppers (*Locustidae*). *Phytopathology*, **18**, 445–448.

Goss, R. W. (1930a). The symptoms of spindle tuber and unmottled curly dwarf of the potato. *Univ. Nebraska Agr. Exp. Sta. Res. Bull.*, **47**, 39 pp.

Goss, R. W. (1930b). Insect transmission of potato-virus diseases. *Phytopathology*, **20**, 136.

Goss, R. W. (1931). Infection experiments with spindle tuber and unmottled curly dwarf of the potato. *Univ. Nebraska Agr. Exp. Sta. Res. Bull.*, **53**, 36 pp.

Goss, R. W., and Peltier, G. L. (1925). Further studies on the effect of environment on potato degeneration diseases. *Univ. Nebraska Agr. Exp. Sta. Res. Bull.*, **29**, 32. pp.

Gralla, J., Steitz, J. A., and Crothers, D. M. 1974. Direct physical evidence for secondary structure in an isolated fragment of R 17 bacteriophage mRNA. *Nature*, **248**, 204–208.

Granboulan, N., and Scherrer, K. (1969). Visualization in the electron microscope and size of RNA from animal cells. *Eur. J. Biochem.*, **9**, 1–20.

Gratz, L. O., and Schultz, E. S. (1931). Observations on certain virus diseases of potatoes in Florida and Maine. *Am. Potato J.*, **7**, 187–200.

Grill, L. K., and Semancik, J. S. (1978). RNA sequences complementary to citrus exocortis viroid in nucleic acid preparations from infected *Gynura aurantiaca*. *Proc. Natl. Acad. Sci. U.S.A.*, **75**, 896–900.

Gross, H. J., Domdey, H., and Sänger, H. L. (1977). Comparative oligonucleotide fingerprints of three plant viroids. *Nucleic Acids Res.*, **4**, 2021–2028.

Gross, H. J., Domdey, H., Lossow, C., Jank, P. Raba, M., Alberty, H., and Sänger, H. L. (1978). Nucleotide sequence and secondary structure of potato spindle tuber viroid. *Nature*, **273**, 203–208.

Hadidi, A., and Diener, T. O. (1977). *De novo* synthesis of potato spindle tuber viroid as measured by incorporation of ^{32}P. *Virology*, **78**, 99–107.

Hadidi, A., and Diener, T. O. (1978). *In vivo* synthesis of potato spindle tuber viroid: kinetic relationship between the circular and linear forms. *Virology*, **86**, 57–65.

Hadidi, A., and Keith, J. M. (1978). *In vivo* synthesis and persistence of potato spindle tuber viroid. *J. Supramolec. Structure, Suppl.*, **2**, 280.

Hadidi, A., Jones, D. M, Gillespie D. H., Wong-Staal, F., and Diener, T. O. (1976). Hybridization of potato spindle tuber viroid to cellular DNA of normal plants. *Proc. Natl. Acad. Sci. U.S.A.*, **73**, 2453–2457.

Hadidi, A., Diener, T. O., and Modak, M. J. (1977). Synthesis of DNA transcripts of potato spindle tuber viroid. *FEBS Letters*, **75**, 123–127.

Hall, T. C., Wepprich, R. K., Davies, J. W., Weathers, L. G., and Semancik, J. S. (1974). Functional distinctions between the ribonucleic acids from citrus exocortis viroid and plant viruses: Cell-free translation and aminoacylation reactions. *Virology*, **61**, 486–492.

Hanneman, R. E., and Singh, R. P. (1972). Seed production in the virus indicator plant *Scopolia sinensis*. *Can. Plant Dis. Surv.*, **52**, 60–61.

Henco, K., Riesner, D., and Sänger, H. L. (1977). Conformation of viroids. *Nucleic Acids Res.*, **4**, 177–194.

Hollings, M., and Stone, O. M. (1970). Attempts to eliminate chrysanthemum stunt from chrysanthemum by meristem-tip culture after heat-treatment. *Ann. Appl. Biol.*, **65**, 311–315.

Hollings, M., and Stone, O. M. (1973). Some properties of chrysanthemum stunt, a virus with the characteristics of an uncoated ribonucleic acid. *Ann. Appl. Biol.*, **74**, 333–348.

Holmes, F. O. (1961a). Does cadang-cadang disease spread from diseased to healthy coconut trees? *FAO Plant Protect. Bull.*, **9**, 139–143.

Holmes, F. O. (1961b). Correlation of cadang-cadang disease in coconut and certain diseases in weeds. *FAO Plant Protect. Bull.*, **9**, 148–152.

Horst, R. K. (1975). Detection of a latent infectious agent that protects against infection by chrysanthemum chlorotic mottle viroid. *Phytopathology*, **65**, 1000–1003.

Horst, R. K., and Romaine, C. P. (1975). Chrysanthemum chlorotic mottle: A viroid disease. *New York's Food and Life Sciences Quarterly*, **8**, 11–14.

Horst, R. K., Langhans, R. W., and Smith, S. H. (1977). Effects of chrysanthemum stunt, chlorotic mottle, aspermy and mosaic on flowering and rooting of chrysanthemums. *Phytopathology*, **67**, 9–14.

Hunter, D. E., Darling, D. H., and Beale, W. L. (1969a). Seed transmission of potato spindle tuber virus. *Am. Potato J.*, **46**, 247–250.

Hunter, G. D. (1974). Scrapie. *Progr. Med. Virol.*, **18**, 289–306.

Hunter, G. D., and Millson, G. C. (1967). Attempts to release the scrapie agent from tissue debris. *J. Comp. Path.*, **77**, 301–307.

Hunter, G. D., Kimberlin, R. H., and Gibbons, R. A. (1968). Scrapie: A modified membrane hypothesis. *J. Theoret. Biol.*, **20**, 355–357.

Hunter, G. D., Gibbons, R. A., Kimberlin, R. H., and Millson, G. C. (1969b). Further studies of the infectivity and stability of extracts and homogenates derived from scrapie affected mouse brains. *J. Comp. Path.*, **79**, 101–108.

Hunter, G. D., Kimberlin, R. H., Collis, S., and Millson, G. C. (1973). Viral and non-viral properties of the scrapie agent. *Ann clin. Res.*, **5**, 262–267.

Hunter, G. D., Collis, S. C., Millson, G. C., and Kimberlin, R. H. (1976). Search for scrapie-specific RNA and attempts to detect an infectious DNA or RNA. *J. Gen. Virol.*, **32**, 157–162.

Hunter, J. E., and Rich, A. E. (1964a). The effect of potato spindle tuber virus on growth and yield of Saco potatoes. *Am. Potato J.*, **41**, 113–116.

Hunter, J. E., and Rich, A. E. (1964b). Liquid nitrogen as an aid in studying potato spindle tuber virus. *Phytopathology*, **54**, 488.

Ikegami, M., and Fraenkel-Conrat, H. (1978). RNA-dependent RNA polymerase of tobacco plants. *Proc. Natl. Acad. Sci. U.S.A.*, **75**, 2122–2124.

Inman, R. B., and Schnös, M. (1970). Partial denaturation of thymine- and 5-bromouracil-containing λ DNA in alkali. *J. Mol. Biol.*, **49**, 93–98.

Jelinek, W., and Darnell, J. E. (1972). Double-stranded regions in heterogeneous nuclear RNA from HeLa cells. *Proc. Natl. Acad. Sci. U.S.A.*, **69**, 2537–2541.

Kacian, D. L., Mills, D. R., Kramer, F. R., and Spiegelman, S. (1972). A replicating RNA molecule suitable for a detailed analysis of extracellular evolution and replication. *Proc. Natl. Acad. Sci. U.S.A.*, **69**, 3038–3042.

Kanehisa, T., Fujitani, H., Sano, M., and Tanaka, T. (1971). Studies on low molecular weight RNA of chromatin. Effects on template activity of chick liver chromatin. *Biochim. Biophys. Acta*, **240**, 46–55.

Kanehisa, T., Tanaka, T., and Kano, Y. (1972). Low molecular RNA associated with chromatin: Purification and characterization of RNA that stimulates RNA synthesis. *Biochim. Biophys. Acta*, **277**, 584–589.

Kanehisa, T., Oki, Y., and Ikuta, K. (1974). Partial specificity of low-molecular-weight RNA that stimulates RNA synthesis in various tissues. *Arch. Biochem. Biophys.*, **165**, 146–152.

Kanehisa, T., Kitazume, Y., Ikuta, K., and Tanaka, Y. (1977a). Release of template restriction in chromatin by nuclear 4.5 S RNA. *Biochim. Biophys. Acta*, **475**, 501–513.

Kanehisa, T., Kitazume, Y., and Matsui, M. (1977b). Interaction of chromatin components with nuclear 5.0 S RNA fraction that stimulates RNA synthesis. *Biochim. Biophys. Acta*, **479**, 265–278.

Kaper, J. M., Tousignant, M. E., and Lot, H. (1976). A low molecular weight replicating RNA associated with a divided genome plant virus: Defective or satellite RNA? *Biochem. Biophys. Res. Commun.*, **72**, 1237–1243.

Keller, J. R. (1951). Report on indicator plants for chrysanthemum stunt virus and on a previously unreported chrysanthemum virus. *Phytopathology*, **41**, 947–949.

Keller, J. R. (1953). Investigations on chrysanthemum stunt virus and chrysanthemum virus Q. *Cornell Univ. Agr. Exp. Sta. Memoir*, **324**, 40 pp.

Kent, G. C. (1953). Cadang-cadang of coconut. *Philippine Argiculturist,* **37**, 228–240.

Kimberlin, R. H. (1976). Experimental scrapie in the mouse: A review of an important model disease. *Science Progress Oxford*, **63**, 461–481.

Kimberlin, R. H., and Marsh, R. F. (1975). Comparison of scrapie and transmissible mink encephalopathy in hamsters. I. Biochemical studies of brain during development of disease. *J. Infect. Disease*, **131**, 97–103.

Kleinschmidt, A. K., and Zahn, R. K. (1959). Ueber Desoxyribonucleinsäure-Molekeln in Protein-Mischfilmen. *Z. Naturforsch. B.*, **14**, 770–779.

Klump, H., Riesner, D., and Sänger, H. D. (1978). Calorimetric studies on viroids. *Nucleic Acids Res.*, **5**, 1581–1587.

Knorr, L. C., and Reitz, H. J. (1959). Exocortis in Florida. In: *Citrus Virus Diseases*, Proc. Conf. on Citrus Virus Diseases, Univ. Calif. Div. Agr. Sci., p. 141–150.

Knorr, L. C., DuCharme, E. P., and Banfi, A. (1951). La exocortis en los montes citricos de la Argentina. *IDIA (Inform. Direcc. Invest. Agr. B. Aires)* **4**, 8–12 (quoted in Knorr and Reitz, 1959).

Kühl, L. (1964). Isolation of plant nuclei. *Z. Naturforsch. B.*, **19**, 525–532.

Laird, E. F., Harjung, M. K., and Weathers, L. G. (1969). Attempts to transmit citrus exocortis by insects. *Plant Dis. Reporter*, **53**, 850–851.

Lang, D. (1970). Molecular weights of coliphages and coliphage DNA. III. Contour length and molecular weight of DNA from bacteriophages T_4, T_5, and T_7, and from bovine papilloma virus. *J. Mol. Biol.*, **54**, 557–565.

Langowski, J., Henco, K., Riesner, D., and Sänger, H. L. (1978). Common structural features of different viroids: Serial arrangement of double helical sections and internal loops. *Nucleic Acids Res.*, **5**, 1589–1610.

Latarjet, R., Muel, B., Haig, D. A., Clarke, M. C., and Alper, T. (1970). Inactivation of the scrapie agent by near monochromatic ultraviolet light. *Nature*, **227**, 1341–1343.

Lawson, R. H. (1968a). Some properties of chrysanthemum stunt virus. *Phytopathology*, **58**, 885.

Lawson, R. H. (1968b). Cineraria varieties as starch lesion test plants for chrysanthemum stunt virus. *Phytopathology*, **58**, 690–695.

LeClerg, E. L., Lombard, P. M., Eddins, A. H., Cook, H. T., and Campbell, J. C. (1944). Effect of different amounts of spindle tuber and leaf roll on yields of Irish potatoes. *Am. Potato J.*, **21**, 60–71.

Lee, C. R., and Singh, R. P. (1972). Enhancement of diagnostic symptoms of potato spindle tuber virus by manganese. *Phytopathology*, **62**, 516–520.

Lee-Huang, S., Sierra, J. M., Naranjo, R., Filipowicz, W., and Ochoa, S. (1977). Eucaryotic oligonucleotides affecting mRNA translation. *Arch. Bichem. Biophys.*, **180**, 276–287.

Leont'eva, Yu. A. (1964). Identification of potato gothic. *Izv. Kuibyshev. sel.-Khoz. Inst.*, **14**, 279–286 (quoted from *Rev. Appl. Mycol.*, **45**, 273, 1966).

Le Roy, C., Stussi-Garaud, C., and Hirth, L. (1977). RNA-dependent RNA polymerases in uninfected and in alfalfa mosaic virus-infected tobacco plants. *Virology*, **82**, 48–62.

Lewandowski, L. J., Kimball, P. C., and Knight, C. A. (1971). Separation of the infectious ribonucleic acid of potato spindle tuber virus from double-stranded ribonucleic acid of plant tissue extracts. *J. Virol.*, **8**, 809–812.

Ling, P. (1972). A report on exocortis virus in Taiwan. In: *Proc. 5th Conf. Intl. Org. Citrus Virologists* (W. C. Price, Ed.), Univ. Florida Press, Gainesville, pp. 102–104.

Lister, R. M., and Hadidi, A. F. (1971). Some properties of apple chlorotic leaf spot virus and their relation to purification problems. *Virology*, **45**, 240–251.

Lockhart, B. E. L., and Semancik, J. S. (1968). Inhibition of the multiplication of a plant virus by actinomycin D. *Virology*, **36**, 504–506.

Lockhart, B. E. L., and Semancik, J. S. (1969). Differential effect of actinomycin D on plant-virus multiplication. *Virology*, **39**, 362–365.

Loening, U. E. (1967). The fractionation of high-molecular-weight ribonucleic acid by polyacrylamide-gel electrophoresis. *Biochem. J.*, **102**, 251–257.

MacLachlan, D. S. (1960). Potato spindle tuber in Eastern Canada. *Am. Potato J.*, **37**, 13–17.

Manzer, F. E., and Merriam, D. (1961). Field transmission of the potato spindle tuber virus and virus X by cultivating and hilling equipment. *Am. Potato J.*, **38**, 346–352.

Manzer, F. E., Akeley, R. V., and Merriam, D. (1964). Resistance in *Solanum tuberosum* to mechanical inoculation with the potato spindle tuber virus. *Am. Potato J.*, **41**, 411–416.

Marsh, R. F., and Hanson, R. P. (1969). Physical and chemical properties of the transmissible mink encephalopathy agent. *J. Virol.*, **3**, 176–180.

Marsh, R. F., and Kimberlin, R. H. (1975). Comparison of scrapie and transmissible mink encephalopathy in hamsters. II. Clinical signs, pathology, and pathogenesis. *J. Inf. Disease*, **131**, 104–110.

Marsh, R. F., Semancik, J. S., Medappa, K. C., Hanson, R. P., and Rueckert, R. R. (1974). Scrapie and transmissible mink encephalopathy: Search for infectious nucleic acid. *J. Virol.*, **13**, 993–996.

Marsh, R. F., Malone, T. G., Semancik, J. S., Lancaster, W. D., and Hanson, R. P. (1978). Evidence for an essential DNA component in the scrapie agent. *Nature*, **275**, 146–147.

Martin, W. H. (1922). "Spindle tuber," a new potato trouble. *Hints to Potato Growers, N.J. State Potato Assoc.*, **3**, No. 8, 4 pp.

McClean, A. P. D. (1931). Bunchy top disease of tomato. *S. Africa Dept. Agr. Sci. Bull.*, **100**, 36 pp.

McClean, A. P. D. (1935). Further investigations on the bunchy top disease of tomato. *S. Africa Dept. Agr. Sci. Bull.*, **139**, 46 pp.

McClean, A. P. D. (1948). Bunchy-top disease of the tomato: Additional host plants, and the transmission of the virus through the seed of infected plants. *S. Africa Dept. Agr. Sci. Bull.*, **256**, 28 pp.

McClean, A. P. D. (1950). Virus infections of citrus in South Africa. *Farming in S. Africa*, **25**, 261–262, 289–296 (quoted in Knorr and Reitz, 1959).

McClements, W. (1975). Electron microscopy of RNA: Examination of viroids and a method for mapping single-stranded RNA. Ph.D. Thesis, Univ. Wisconsin, Madison.

McClements, W. L., and Kaesberg, P. (1977). Size and secondary structure of potato spindle tuber viroid. *Virology*, **76**, 477–484.

McKinney, H. H. (1973). Comments on virus terminology and the "viroid." *Phytopathology*, **63**, 438.

Merriam, D., and Bonde, R. (1954). Dissemination of spindle tuber by contaminated tractor wheels and by foliage contact with diseased potato plants. *Phytopathology*, **44**, 111.

Mikoshiba, K., Tsukada, Y., Haruna, I., and Watanabe, I. (1974). RNA-dependent RNA synthesis in rat brain. *Nature*, **249**, 445–448.

Mills, D. R., Peterson, R. L., and Spiegelman, S. (1967). An extracellular Darwinian experiment with a self-duplicating nucleic acid molecule. *Proc. Natl. Acad. Sci. U.S.A.*, **58**, 217–224.

Mills, D. R., Kramer, F. R., and Spiegelman, S. (1973). Complete nucleotide sequence of a replicating RNA molecule. *Science*, **180**, 916–927.

Mills, D. R., Kramer, F. R., Dobkin, C., Nishihara, T., and Spiegelman, S. (1975). Nucleotide sequence of microvariant RNA: Another small replicating molecule. *Proc. Natl. Acad. Sci. U.S.A.*, **72**, 4252–4256.

Montagnier, L. (1968). Présence d'un acide ribonucléique en double chaîne dans des cellules animales. *C. R. Acad. Sci. Ser. D.*, **267**, 1417–1420.

Moreira, S. (1955). A moléstia "exocortis" e o cavalo de limoeiro cravo. *Revista Agricultura (Piracicaba)*, **30**, 99–112.

Moreira, S. (1959). Rangpur lime disease and its relationship to exocortis. In: *Citrus Virus Diseases* (J. M. Wallace, Ed.), Univ. California Div. Agr. Sci., pp. 135–140.

Moreira, S. (1961). A quick field test for exocortis. In: *Proc. 2nd Conf. Intl. Org. Citrus Virologists* (W. C. Price, Ed), Univ. Florida Press, Gainesville pp. 40–42.

Morris, T. J., and Smith, E. M. (1977). Potato spindle tuber disease: Procedures for the detection of viroid RNA and certification of disease-free potato tubers. *Phytopathology*, **67**, 145–150.

Morris, T. J., and Wright, N. S. (1975). Detection on polyacrylamide gel of a diagnostic nucleic acid from tissue infected with potato spindle tuber viroid. *Am. Potato J.*, **52**, 57–63.

Mould, D. L., Dawson, A. M., and Smith, W. (1965). Scrapie in mice. The stability of the agent to various suspending media, pH and solvent extraction. *Res. Vet. Sci.*, **6**, 151–154.

Mühlbach, H. P., and Sänger, H. L. (1977). Multiplication of cucumber pale fruit viroid in inoculated tomato leaf protoplasts. *J. Gen. Virol.*, **35**, 377–386.

Mühlbach, H. P., Camacho-Henriquez, A., and Sänger, H. L. (1977). Isolation and properties of protoplasts from leaves of healthy and viroid-infected tomtato plants. *Plant Science Letters*, **8**, 183–189.

Mushin, R. (1942). Serological studies on plant viruses. *Austr. J. Exp. Biol. Med. Sci.*, **20**, 59–63.

Niblett, C. L., Hedgcoth, C., and Diener, T. O. (1976). Base composition of potato spindle tuber viroid. *Beltsville Symp. Virol. in Agric.*, Abstracts, p. 27.

Noordam, D. (1952). Virusziekten bij Chrysanten in Nederland. *Tijdschr. Pl. Ziekt.*, **58**, 121–189.

Norman, G. G. (1965). The incidence of exocortis virus in Florida citrus varieties. In: *Proc. 3rd Intl. Org. Citrus Virologists* (W. C. Price, Ed.), Univ. Florida Press, Gainesville, pp. 124–127.

Notkins, A. L. (1965). Recovery of an infectious ribonucleic acid from the lactic dehydrogenase virus following extraction with butanol or chloroform. *Biochim. Biophys. Acta*, **103**, 509–511.

Öberg, B. (1970). Biochemical and biological characteristics of carbethoxylated polio virus and viral RNA. *Biochim. Biophys. Acta*, **204**, 430–440.

Öberg, B., and Philipson, L. (1972). Binding of histidine to tobacco mosaic virus RNA. *Biochem. Biophys. Res. Commun.*, **48**, 927–932.

O'Brien, M. J. (1972). Hosts of potato spindle tuber virus in suborder Solanineae. *Am. Potato J.*, **49**, 70–72.

O'Brien, M. J., and Raymer, W. B. (1964). Symptomless hosts of the potato spindle tuber virus. *Phytopathology*, **54**, 1045–1047.

Ocfemia, G. O. (1937). The probable nature of "cadang-cadang' disease of coconut. *Philippine Agriculturist*, **26**, 338–340.

Ocefemia, G. O. (1950). Measures suggested for the control of the infectious form of cadang-cadang disease of coconut. *College of Agriculture Biweekly Bull.*, **15** (1), 1–2 (quoted by Kent, 1953).

Ohe, K. (1972). Virus-coded origin of a low-molecular-weight RNA from KB cells infected with adenovirus 2. *Virology*, **47**, 726–733.

Ohe, K., and Weissman, S. M. (1971). The nucleotide sequence of a low molecular weight ribonucleic acid from cells infected with adenovirus 2. *J. Biol. Chem.*, **246**, 6991–7009.

Olson, C. J. (1949). A preliminary report on transmission of chrysanthemum stunt. *Chrys. Soc. Amer. Bull.*, **17**, 2–9 (quoted in Keller, 1953).

Olson, E. O. (1968). Review of recent research on exocortis disease. In: *Proc. 4th Conf. Intl. Org. Citrus Virologists* (J. F. L. Childs, Ed.), Univ. Florida Press, Gainesville, pp. 92–96.

Olson, E. O., and Shull, A. V. (1956). Exocortis and xyloporosis—Bud transmission virus diseases of Rangpur and other mandarin-lime rootstocks. *Plant Dis. Reporter*, **40**, 939–946.

Orton, W. A. (1914). Potato wilt, leaf-roll, and related diseases. *U.S. Dept. Agr. Bull.*, **64**, 48 pp.

Owens, R. A. (1978). *In vitro* synthesis and characterization of DNA complementary to potato spindle tuber viroid. *Virology*, **89**, 380–387.

Owens, R. A., Erbe, E., Hadidi, A., Steere, R. L., and Diener, T. O. (1977). Separation and infectivity of circular and linear forms of potato spindle tuber viroid. *Proc. Natl. Acad. Sci. U.S.A.*, **74**, 3859–3863.

Owens, R. A., Smith, D. R., and Diener, T. O. (1978). Measurement of viroid sequence homology by hybridization with complementary DNA prepared *in vitro*. *Virology*, **89**, 388–394.

Pattison, I. H., and Jones, K. M. (1967). The possible nature of the transmissible agent of scrapie. *Vet. Record*, **80**, 2–9.

Peacock, A. C., and Dingman, C. W. (1967). Resolution of multiple ribonucleic acid species by polyacrylamide gel electrophoresis. *Biochemistry*, **6**, 1818–1827.

Philipson, L., and Lindberg, U. (1974). Reproduction of adenoviruses. In: *Comprehensive Virology* (H. Fraenkel-Conrat and R. R. Wagner, Eds.), Plenum Press, New York, Vol. 3, pp. 143–227.

Planes, S., Marti-Fabrigat, F., Fuertes, C., Garcia, J., and Aparicio, M. (1968). Exocortis in the citrus area of Valencia. In: *Proc. 4th Conf. Intl. Org. Citrus Viorologists* (J. F. L. Childs, Ed.), Univ. Florida Press, Gainesville, pp. 100–101.

Prensky, W. (1975). The radioiodination of RNA and DNA to high specific activities. In: *Methods in Cell Biology* (D. Prescott, Ed.), Academic Press, New York, Vol. 13, pp. 121–151.

Price, W. C. (1971). Cadang-cadang of coconut—A review. *Plant Science*, **3**, 1–13.

Price, W. C., and Bigornia, A. E. (1969). Further studies on the epidemiology of coconut cadang-cadang disease. *FAO Plant Prot. Bull.*, **17**, 11–16.

Price, W. C., and Bigornia, A. E. (1972). Evidence for spread of cadang-cadang disease of coconut from tree to tree. *FAO Plant Prot. Bull.*, **20**, 33–135.

Ralph, R. K., and Bellamy, A. R. (1964). Isolation and purification of undegraded ribonucleic acid. *Biochim. Biophys. Acta*, **87**, 9–16.

Ralph, R. K., Matthews, R. E. F., Matus, A. I., and Mandel, H. G. (1965). Isolation and properties of double-stranded viral RNA from virus-infected plants. *J. Mol. Biol.*, **11**, 202–212.

Randerath, E., Yu, C. T., and Randerath, K. (1972). Base analysis of ribopolynucleotides by chemical tritium labelling: A methodological study with model nucleosides and purified tRNA species. *Anal. Biochem.*, **48**, 172–198.

Randles, J. W. (1975). Association of two ribonucleic acid species with cadang-cadang disease of coconut palm. *Phytopathology*, **65**, 163–167.

Randles, J. W., Rillo, E. P., and Diener, T. O. (1976). The viroidlike structure and cellular location of anomalous RNA associated with the cadang-cadang disease. *Virology*, **74**, 128–139.

Randles, J. W., Boccardo, G., Retuerma, M. L., and Rillo, E. P. (1977). Transmission of the RNA species associated with cadang-cadang of coconut palm, and the insensitivity of the disease to antibiotics. *Phytopathology*, **67**, 1211–1216.

Raymer, W. B., and Diener, T. O. (1969). Potato spindle tuber virus: A plant virus with properties of a free nucleic acid. I. Assay, extraction, and concentration. *Virology*, **37**, 343–350.

Raymer, W. B., and O'Brien, M.J. (1962). Transmission of potato spindle tuber virus to tomato. *Am. Potato J.*, **39**, 401–408.

Raymer, W. B., O'Brien, M. J., and Merriam, D. (1964). Tomato as a source of and indicator plant for the potato spindle tuber virus. *Am. Potato J.*, **41**, 311–314.

Reanney, D. C. (1975). A regulatory role for viral RNA in eukaryotes. *J. Theoret. Biol.*, **49**, 461–492.

Reanney, D. (1976). Extrachromosomal elements as possible agents of adaptation and development. *Bacteriol. Rev.*, **40**, 552–590.

Reinking, O. A. (1950). Preliminary report on the cadang-cadang disease and on soil deficiency troubles of coconuts in the Philippines. *Plant Dis. Reporter*, **34**, 300–303.

Reitz, H. J., and Knorr, L. C. (1957). Occurrence of Rangpur lime disease in Florida and its concurrence with exocortis. *Plant Dis. Reporter*, **41**, 235–240.

Reunova, G. D., Reunov, A. V., and Reifman, V. G. (1973). Effect of actinomycin D on potato virus X multiplication in *Datura stramonium* leaves. *Virology*, **52**, 502–506.

Robertson, H. D., and Hunter, T. (1975). Sensitive methods for the detection and characterization of double helical ribonucleic acid. *J. Biol. Chem.*, **250**, 418–425.

Ro-Choi, T. S., and Busch, H. (1974). Low molecular weight nuclear RNA. In: *The Molecular Biology of Cancer* (H. Busch, Ed.), Academic Press, New York, pp. 241–276.

Ro-Choi, T. S., Reddy, R., Henning, D., Takano, T., Taylor, C. W., and Busch, H.

(1972). Nucleotide sequence of 4.5 S ribonucleic acid I of Novikoff hepatoma cell nuclei. *J. Biol. Chem.*, **247**, 3205–3222.

Roistacher, C. N., Calavan, E. C., and Blue, R. L. (1969). Citrus exocortis virus— Chemical inactivation on tools, tolerance to heat and separation of isolates. *Plant Dis. Reporter*, **53**, 333-336.

Romaine, C. P., and Horst, R. K. (1974). Evidence suggesting a viroid etiology for chrysanthemum chlorotic mottle disease. *Proc., Am. Phytopath. Soc.*, **1**, 143.

Romaine, C. P., and Horst, R. K. (1975). Suggested viroid etiology for chrysanthemum chlorotic mottle disease. *Virology*, **64**, 86–95.

Romaine, C. P., and Zaitlin, M. (1978). RNA-dependent RNA polymerases in uninfected and tobacco mosaic virus-infected tobacco leaves: Viral-induced stimulation of a host polymerase activity. *Virology*, **86**, 241–253.

Romero, J. (1972). RNA synthesis in broadbean leaves infected with broadbean mottle virus. *Virology*, **48**, 591–594.

Rossetti, V. (1961). Testing for exocortis. In: *Proc. 2nd Conf. Intl. Org. Citrus Virologists* (W. C. Price, Ed.), Univ. Florida Press Gainesville, pp. 43–49.

Sakai, F., and Takebe, I. (1974). Protein synthesis in tobacco mesophyll protoplasts induced by tobacco mosaic virus infection. *Virology*, **62**, 426–433.

Salibe, A. A., and Moreira, S. (1965a). New test varieties for exocortis virus. In: *Proc. 3rd Conf. Intl. Org. Citrus Virologists* (W. C. Price, Ed.), Univ. Florida Press, Gainesville, pp. 119–123.

Salibe, A. A., and Moreira, S. (1965b). Seed transmission of exocortis virus. In: *Proc. 3rd Conf. Intl. Org. Citrus Virologists* (W. C. Price, Ed.), Univ. Florida Press, Gainesville, pp. 139–142.

Sambrook, J. (1977). Adenovirus amazes at Cold Spring Harbor. *Nature*, **268**, 101–104.

Samson, R. W. (1927). A study of the properties and nature of the virus of the spindle-tuber disease of potatoes. M. S. Thesis, Univ. Nebraska, Lincoln.

Sänger, H. L. (1972). An infectious and replicating RNA of low molecular weight: The agent of the exocortis disease of citrus. *Adv. Bioscience*, **8**, 103–116.

Sänger, H. L., and Brandenburg, E. (1961). Ueber die Gewinnung von infektiösem Pressaft aus "Wintertyp"—Pflanzen des Tabak-Rattle-Virus durch Phenolextraktion. *Naturwiss.*, **48**, 391.

Sänger, H. L., and Knight, C. A. (1963). Action of actinomycin D on RNA synthesis in healthy and virus-infected tobacco leaves. *Biochem. Biophys. Res. Commun.*, **13**, 455–461.

Sänger, H. L., and Ramm, K. (1975). Radioactive labelling of viroid-RNA. In: *Modification of the Information Content of Plant Cells* (R. Markham, D. R. Davies, D. A. Hopwood, and R. W. Horne, Eds.), North-Holland/American Elsevier Publ. Co., Amsterdam, pp. 229–252.

Sänger, H. L., Klotz, G., Riesner, D., Gross, H. J., and Kleinschmidt, A. K. (1976). Viroids are single-stranded covalently closed circular RNA molecules existing as highly base-paired rod-like structures. *Proc. Natl. Acad. Sci. U.S.A.*, **73**, 3852–3856.

Sasaki, M., and Shikata, E. (1977a). Studies on the host range of hop stunt disease in Japan. *Proc. Japan Acad.*, **53B**, 103–108.

Sasaki, M., and Shikata, E. (1977b). On some properties of hop stunt disease agent, a viroid. *Proc. Japan Acad.*, **53B**, 109–112.

Schneider, I. R. (1971). Characteristics of a satellite-like virus of tobacco ringspot virus. *Virology,* **45,** 108–122.

Schultz, E. S., and Folsom, D. (1923a). Transmission, variation, and control of certain degeneration diseases of Irish potatoes. *J. Agr. Res.,* **25,** 43–117.

Schultz, E. S., and Folsom, D. (1923b). Spindling-tuber and other degeneration diseases of Irish potatoes. *Phytopathology,* **13,** 40.

Schultz, E. S., and Folsom, D. (1923c). A "spindling-tuber disease" of Irish potatoes. *Science,* **57,** 149.

Schultz, E. S., and Folsom D. (1925). Infection and dissemination experiments with degeneration diseases of potatoes. Observations in 1923. *J. Agr. Res.,* **30,** 493–528.

Schwartz, E. F., and Stollar, B. D. (1969). Antibodies to polyadenylate-polyuridylate copolymers as reagents for double-stranded RNA and DNA:RNA hybrid complexes. *Biochem. Biophys. Res. Commun.,* **35,** 115–120.

Semal, J. (1966). Effects of actinomycin D in plant virology. *Phytopath. Z.,* **59,** 55–71.

Semancik, J. S. (1974). Detection of polyadenylic acid sequences in plant pathogenic RNAs. *Virology,* **62,** 288–291.

Semancik, J. S., and Geelen, J. L. M. C. (1975). Detection of DNA complementary to pathogenic viroid RNA in exocortis disease. *Nature,* **256,** 753–756.

Semancik, J. S., and Vanderwoude, W. J. (1976). Exocortis viroid: Cytopathic effects at the plasma membrane in association with pathogenic RNA. *Virology,* **69,** 719–726.

Semancik, J. S., and Weathers, L. G. (1968a). Characterization of infectious nucleic acid associated with infection by exocortis virus of citrus. *Phytopathology,* **58,** 1067.

Semancik, J. S., and Weathers, L. G. (1968b). Exocortis virus of citrus: Association of infectivity with nucleic acid preparations. *Virology,* **36,** 326–328.

Semancik, J. S., and Weathers, L. G. (1970a). Properties of the infectious forms of exocortis virus of citrus. *Phytopathology,* **60,** 732–736.

Semancik, J. S., and Weathers, L. G. (1970b). Multiple infectious forms of a free-RNA plant virus. *Phytopathology,* **60,** 1313.

Semancik, J. S., and Weathers, L. G. (1971). Evidence for an infectious double-stranded RNA plant virus. *Phytopathology,* **61,** 910.

Semancik, J. S., and Weathers, L. G. (1972a). Exocortis virus: An infectious free-nucleic acid plant virus with unusual properties. *Virology,* **47,** 456–466.

Semancik, J. S., and Weathers, L. G. (1972b). Exocortis disease: Evidence for a new species of "infectious" low molecular weight RNA in plants. *Nature New Biology,* **237,** 242–244.

Semancik, J. S., and Weathers, L. G. (1972c). Pathogenic 10 S RNA from exocortis disease recovered from tomato bunchy-top plants similar to potato spindle tuber virus infection. *Virology,* **49,** 622–625.

Semancik, J. S., Magnuson, D. S., and Weathers, L. G. (1973a). Potato spindle tuber disease produced by pathogenic RNA from citrus exocortis disease: Evidence for the identity of the causal agent. *Virology,* **52,** 292–294.

Semancik, J. S., Morris, T. J., and Weathers, L. G. (1973b). Structure and conformation of low molecular weight pathogenic RNA from exocortis disease. *Virology,* **53,** 448–456.

Semancik, J. S., Morris, T. J., Weathers, L. G., Rodorf, B. F., and Kearns, D. R. (1975). Physical properties of a minimal infectious RNA (viroid) associated with the exocortis disease. *Virology,* **63,** 160–167.

Semancik, J. S., Tsuruda, D., Zaner, L., Geelen, J. L. M. C., and Weathers, L. G. (1976). Exocortis disease: Subcellular distribution of pathogenic (viroid) RNA. *Virology,* **69,** 669–676.

Semancik, J. S., Conejero, V., and Gerhart, J. (1977). Citrus exocortis viroid: Survey of protein synthesis in *Xenopus laevis* oocytes following addition of viroid RNA. *Virology,* **80,** 218–221.

Siegel, A., Zaitlin, M., and Sehgal, O. P. (1962). The isolation of defective tobacco mosaic virus strains. *Proc. Natl. Acad. Sci. U.S.A.,* **48,** 1845–1851.

Siegel, A., Hari, V., Montgomery, I., and Kolacz, K. (1976). A messenger RNA for capsid protein isolated from tobacco mosaic virus-infected tissue. *Virology,* **73,** 363–371.

Sill, W. H., Jr., Bigornia, A. E., and Pacumbaba, R. P. (1963). Incidence of cadang-cadang in coconut trees of different ages. *FAO Plant Protect. Bull.,* **11,** 49–58.

Sinclair, J. B., and Brown, R. T. (1960). Effect of exocortis disease on four citrus rootstocks. *Plant Dis. Reporter,* **44,** 180–183.

Singer, B., and Fraenkel-Conrat, H. (1963). Studies of nucleotide sequences in TMV-RNA. I. Stepwise use of phosphodiesterase. *Biochim. Biophys. Acta,* **72,** 534–543.

Singh, A., and Sänger, H. L. (1976). Chromatographic behaviour of the viroids of the exocortis disease of citrus and the spindle tuber disease of potato. *Phytopath. Z.,* **87,** 143–160.

Singh, R. P. (1970a). Occurrence, symptoms, and diagnostic hosts of strains of potato spindle tuber virus. *Phytopathology,* **60,** 1314.

Singh, R. P. (1970b). Seed transmission of potato spindle tuber virus in tomato and potato. *Am. Potato J.,* **47,** 225–227.

Singh, R. P. (1971). A local lesion host for potato spindle tuber virus. *Phytopathology,* **61,** 1034–1035.

Singh, R. P. (1973). Experimental host range of the potato spindle tuber "virus." *Am. Potato J.,* **50,** 111–123.

Singh, R. P. (1977). Piperonyl butoxyde as a protectant against potato spindle tuber viroid infection. *Phytopathology,* **67,** 933–935.

Singh, R. P., and Bagnall, R. H. (1968). Infectious nucleic acid from host tissues infected with potato spindle tuber virus. *Phytopathology,* **58,** 696–699.

Singh, R. P., and Clark, M. C. (1971a). Infectious low molecular weight ribonucleic acid from tomato infected with potato spindle tuber virus. *Phytopathology,* **61,** 911.

Singh, R. P., and Clark, M. C. (1971b). Infectious low-molecular-weight ribonucleic acid. *Biochem. Biophys. Res. Commun.,* **44,** 1077–1083.

Singh, R. P., and Clark, M. C. (1973a). Nature of the 3′-terminus and molecular size of spindle tuber metavirus. *Abstracts, Second Intl. Congr. Plant Pathology, Minneapolis, Minn.,* Abstract No. 282.

Singh, R. P., and Clark, M. C. (1973b). Similarity of host response to both potato spindle tuber and citrus exocortis viruses. *FAO Plant Protect. Bull.,* **21,** 121–125.

Singh, R. P., and Finnie, R. E. (1977). Stability of potato spindle tuber viroid in freeze-dried leaf powder. *Phytopathology,* **67,** 283–286.

Singh, R. P., and O'Brien, M. J. (1970). Additional indicator plants for potato spindle tuber virus. *Am. Potato J.,* **47,** 367–371.

Singh, R. P., Benson, A. P., and Salama, F. M. (1966). Purification and electron microscopy of potato spindle tuber virus. *Phytopathology,* **56,** 901–902.

Singh, R. P., Finnie, R. E., and Bagnall, R. H. (1970). Relative prevalence of mild and severe strains of potato spindle tuber virus in Eastern Canada. *Am. Potato J.*, **47**, 289–293.

Singh, R. P., Finnie, R. E., and Bagnall, R. H. (1971). Losses due to the potato spindle tuber virus. *Am. Potato J.*, **48**, 262–267.

Singh, R. P., Clark, M. C., and Weathers, L. G. (1972). Similarity of host symptoms induced by citrus exocortis and potato spindle tuber causal agents. *Phytopathology*, **62**, 790.

Singh, R. P., Michniewicz, J. J., and Narang, S. A. (1974). Multiple forms of potato spindle-tuber metavirus ribonucleic acid. *Can. J. Biochem.*, **52**, 809–812.

Singh, R. P., Michniewicz, J. J., and Narang, S. A. (1975a). Isolation of infectious DNA-associated RNA from plants infected with potato spindle tuber metavirus. In: *International Virology*. Abstracts, 3rd Intl. Congr. Virology, Madrid, Spain, p. 234.

Singh, R. P., Michniewicz, J. J., and Narang S. A. (1975b). Isolation of two distinct forms of potato spindle tuber metavirus RNA and their fingerprinting patterns. *Fed. Proceed.*, **34**, 638.

Singh, R. P., Michniewicz, J. J., and Narang, S. A. (1975c). Piperonyl butoxyde, a potent inhibitor of potato spindle tuber viroid in *Scopolia sinensis. Can. J. Biochem.*, **53**, 1130–1132.

Singh, R. P., Michniewicz, J. J., and Narang, S. A. (1976). Separation of potato spindle tuber viroid ribonucleic acid from *Scopolia sinensis* into three infectious forms and the purification and oligonucleotide patterns of fraction II RNA. Part IX. *Can. J. Biochem.*, **54**, 600–608.

Smith, K. M. (1972). *A Textbook of Plant Virus Diseases, Third edition.* Academic Press, New York, p. 178.

Sogo, J. M., Koller, T., and Diener, T. O. (1973). Potato spindle tuber viroid. X. Visualization and size determination by electron microscopy. *Virology*, **55**, 70–80.

Spencer, D., and Wildman, S. G. (1964). The incorporation of amino acids into protein by cell-free extracts from tobacco leaves. *Biochemistry*, **3**, 954–959.

Stace-Smith, R., and Mellor, F. C. (1970). Eradication of potato spindle tuber virus by thermotherapy and axillary bud culture. *Phytopathology*, **60**, 1857–1858.

Stedman, T. L. (1961). *Stedman's Medical Dictionary, Twentieth edition.* Williams & Wilkins Co., Baltimore, Md. 1680 pp.

Steere, R. L. (1955). Concepts and problems concerning the assay of plant viruses. *Phytopathology*, **45**, 196–208.

Stern, R., and Friedman, R. M. (1970). Double-stranded RNA synthesized in animal cells in the presence of actinomycin D. *Nature*, **226**, 612–616.

Stollar, B. D., and Diener, T. O. (1971). Potato spindle tuber viroid. V. Failure of immunological tests to disclose double-stranded RNA or RNA-DNA hybrids. *Virology*, **46**, 168–170.

Stollar, V., and Stollar, B. D. (1970). Immunochemical measurement of double-stranded RNA of uninfected and arbovirus-infected mammalian cells. *Proc. Natl. Acad. Sci. U.S.A.*, **65**, 993–1000.

Studier, F. W. (1973). Analysis of bacteriophage T_7 early RNAs and protein on slab gels. *J. Mol. Biol.*, **79**, 237–248.

Stussi-Garaud, C., Lemius, J., and Fraenkel-Conrat, H. (1977). RNA polymerase from tobacco necrosis virus-infected and uninfected tobacco. II. Properties of the bound and soluble polymerases and the nature of their products. *Virology*, **81**, 224–236.

Takahashi, T., and Diener, T. O. (1975). Potato spindle tuber viroid. XIV. Replication in nuclei isolated from infected leaves. *Virology*, **64**, 106–114.

Tewari, K. K., and Wildman, S. G. (1966). Chloroplast DNA from tobacco leaves. *Science*, **153**, 1269–1271.

Teyssier, D., and Dunez, J. (1971). Le rabougrissement du chrysanthème: symptômes et détection. *Annales de Phytopathologie*, **3**, 65–73.

Van Dorst, H. J. M., and Peters, D. (1974). Some biological observations on pale fruit, a viroid-incited disease of cucumber. *Neth. J. Pl. Path.*, **80**, 85–96.

Van Vloten-Doting, L., Kruseman, J., and Jaspars, E. M. J. (1968). The biological function and mutual dependence of bottom component and top component a of alfalfa mosaic virus. *Virology*, **34**, 728–737.

Vogel, R., Bové, C., and Bové, J. M. (1965). Exocortis in Corsica. In: *Proc. 3rd Conf. Intl. Org. Citrus Virologists* (W. C. Price, Ed.), Univ. Florida Press, Gainesville, pp. 134–138.

Walbot, V., and Dure, L. S. III. (1976). Developmental biochemistry of cotton seed embryogenesis and germination. VII. Characterization of the cotton genome. *J. Mol. Biol.*, **101**, 503–536.

Ward, R. L., Porter, D. D., and Stevens, J. G. (1974). Nature of the scrapie agent: Evidence against a viroid. *J. Virol.*, **14**, 1099–1103.

Weathers, L. G. (1965a). Petunia, an herbaceous host of exocortis virus of citrus. *Phytopathology*, **55**, 1081.

Weathers, L. G. (1965b). Transmission of exocortis virus of citrus by *Cuscuta subinclusa*. *Plant Dis. Reporter*, 49, 189–190.

Weathers, L. G., and Calavan, E. C. (1961). Additional indicator plants for exocortis and evidence for strain differences in the virus. *Phytopathology*, **51**, 262–264.

Weathers, L. G., and Greer, F. C. (1968). Additional herbaceous hosts of the exocortis virus of citrus. *Phytopathology*, **58**, 1071.

Weathers, L. G., and Greer, F. C. (1972). *Gynura* as a host for exocortis virus of citrus. In: *Proc. 5th Conf. Intl. Org. Citrus Virologists* (W. C. Price, Ed.), Univ. Florida Press, Gainesville, pp. 95–98.

Weathers, L. G., and Harjung, M. K. (1964). Transmission of citrus viruses by dodder, *Cuscuta subinclusa*. *Plant Dis. Reporter*, **48**, 102–103.

Weathers, L. G., Harjung, M. K., and Platt, R. G. (1965). Some effects of host nutrition on symptoms of exocortis. In: *Proc. 3rd Conf. Intl. Org. Citrus Virologists* (W. C. Price, Ed.), Univ. Florida Press, Gainesville, pp. 102–107.

Weathers, L. G., Greer, F. C., and Harjung, M. K. (1967). Transmission of exocortis virus of citrus to herbaceous plants. *Plant Dis. Reporter*, **51**, 868–871.

Webb, R. E. (1958). Schultz potato virus collection. *Am. Potato J.*, **35**, 615–619.

Welsh, M. F. (1948). Stunt-mottle virus disease of chrysanthemum. *Scientific Agric. (Agr. Inst. Canada)*, **28**, 422.

Werner, H. O. (1926). The spindle-tuber disease as a factor in seed potato production. *Univ. Nebraska, Agr. Exp. Sta. Res. Bull.*, **32**, 128 pp.

White, J. L., and Murakishi, H. H. (1977). *In vitro* replication of tobacco mosaic virus RNA in tobacco callus cultures: Solubilization of membrane-bound replicase and partial purification. *J. Virol.*, **21**, 484–492.

Wickner, R. B. (1976). Killer of *Saccharomyces cerevisiae*: A double-stranded ribonucleic acid plasmid. *Bacter. Rev.*, **40**, 757–773.

Williams, B. (1977). DNA insertions and gene structure. *Nature*, **270**, 295–297.

Yamada, S., and Tanaka, H. (1972). Damage from exocortis in Japan. In: *Proc. 5th Conf. Intl. Org. Citrus Virologists* (W. C. Price, Ed.), Univ. Florida Press, Gainesville, pp. 99–101.

Yot, P., Pinck, M., Haenni, A., Duranton, H. M., and Chapeville, F. (1970). Valine-specific tRNA-like structure in turnip yellow mosaic virus RNA. *Proc. Natl. Acad. Sci. U.S.A.*, **67**, 1345–1352.

Zaitlin, M., and Beachy, R. N. (1974). The use of protoplasts and separated cells in plant virus research. *Adv. Virus Res.*, **19**, 1–35.

Zaitlin, M., and Hariharasubramanian, V. (1970). Proteins in tobacco mosaic virus- and potato spindle tuber virus-infected plants. *Phytopathology*, **60**, 1537–1538.

Zaitlin, M., and Hariharasubramanian, V. (1972). A gel electrophoretic analysis of proteins from plants infected with tobacco mosaic and potato spindle tuber viruses. *Virology*, **47**, 296–305.

INDEX